GENETIC DISEASE CONTROL

Volume 116, Sage Library of Social Research

RECENT VOLUMES IN SAGE LIBRARY OF SOCIAL RESEARCH

GENETIC DISEASE CONTROL

A Social Psychological Approach

FRED MASSARIK
MICHAEL M. KABACK

Volume 116
SAGE LIBRARY OF
SOCIAL RESEARCH

SAGE PUBLICATIONS Beverly Hills London

For information address:

SAGE Publications, Inc.
275 South Beverly Drive
Beverly Hills, California 90212

SAGE Publications Ltd
28 Banner Street
London EC1Y 8QE, England

Printed in the United States of America

Library of Congress Cataloging in Publication Data

Massarik, Fred.
Genetic disease control.

(Sage library of social research)
Bibliography: p.
Includes index.
1. Genetic counseling–Social aspects.
2. Genetic counseling–Psychological aspects.
I. Kaback, Michael M., joint author. II. Title.
[DNLM: 1. Hereditary diseases–Prevention and control. 2. Genetics, Medical. QZ 50 M414g]
RB155.M35 362.1'96042 80-28219

ISBN 0-8039-1054-1
ISBN 0-8039-1055-X (pbk.)

FIRST PRINTING

CONTENTS

PREFACE

Not every Man or Woman was design'd
To propogate and multiply their kind;

Forbid we rightly the Deform's and Foul
To clothe with ill-shap'd Limbs the heav'nly Soul?

Has not the Poet's Song divinely told
Of Births detested in the days of old?

—Callipaedia, or The Art of
Getting Beautiful Children,
Claudius Quillet (N. Rowe, trans.)
London: A. Bell et al., 1720, p. 21.

Since ancient days, folkways and myths, as well as prescriptive medical lore, have provided cautionary guides to the process of conception and birth. Rooted in the world views current in a given place and age, intertwined with prevailing social values no less than with knowledge (actual or presumed) of medical reality, homilies like the following have had their day:

- A child conceived on the night of the new moon may well come to life grossly deformed.
- Offspring of marriages of Black and white cannot find grace in the eye of God.
- The bleeding disease (hemophilia) is strictly a disease of royalty.

Whether "old wives' tales," quasi-religious injunctions, or commonly held "folk wisdom," some notions of genetic process have been part and parcel of virtually every culture and era of social thought. Revealed systematically in the works of anthropology or discussed informally among pregnant women or marriage partners, what is indeed known about genetics—on the basis of systematic research—and what is commonly assumed to be factual are often dramatically different matters. Even among the presumably well-educated, lack of information and misinformation abound concerning matters of heredity and its consequences.

This is not a "medical" book, at least not in a narrow sense. Rather, it addresses some issues relating to genetics and medicine in the perspectives of the social and behavioral sciences. It provides viewpoints, especially bearing on impact of new genetic disease control methods and research data, as prototype, reporting psychosocial research conducted in connection with the California Tay-Sachs Disease Prevention Program.

In this context, we consider how a specified population perceives genetic disease and its consequences, levels of genetics knowledge, and attitudes toward the diagnosis, treatment, cure, and prevention of genetic illness. Additionally, we examine what these people (primarily Jewish, married, and of reproductive age) think ought to be done in terms of genetics testing and screening, and what they are prepared to do, by way of specific intervention such as abortion, in instances in which a genetic disorder is diagnosed *in utero*. Finally, we consider the hypothesis of "stigma," by studying the attitudes and feelings of couples in which one member has been identified as a carrier of Tay-Sachs disease.

The principal empirical data reported derive primarily from a series of sociopsychological research projects conducted by the University of California, Los Angeles, at its Division of Medical Genetics (Harbor-UCLA Medical Center), by UCLA Behavioral and Organizational Science Faculty

(Graduate School of Management), and by an independent consulting group (Behavioral Science Research Associates), 1975-1980. These empirical data on primarily Jewish populations are presented as both substantive and illustrative, with implications for comparable needed research in other populations and in other settings.

One may ask, "Is there any reason why the topic of genetic disease control is any more relevant now than it had been in the past?" The first chapter addresses this issue.

Chapter 1

NEW VISTAS IN GENETIC DISEASE CONTROL

The once Inevitable no longer is inevitable.

–Anonymous

In times past, beyond adherence to popular prescription for simplistic home remedy, there seemed to be little that people could–or were prepared to–do about their hereditary legacy. Even fairly systematic understanding of certain genetic regularities (such as Mendelian laws, rudimentary knowledge of genes and chromosomes) widely appeared as irrelevant curiosity or as foreordained circumstance, but hardly as something that could be affected in direct manner. "Having that deformed child is a burden that just must be borne." "There will always be those who are mentally retarded–this simply is an affliction that the parents, and society, must endure." And at one time, more drastic views of eugenics, the tragic views of "pure-bred" Aryan master races, the "brave new worlds," these and other ideologies illustrate reactions at the other extreme: the proposed creation of vast systems of centralized effort–in the vein of some kind of giant "societal engineering"–that would alter a people's genetic character, and lead to some imagined never-never land of total genetic health.

At another point in the spectrum, there were those who looked beyond the issue of genetics and denied all relevance of any inherent predisposition. Most articulately, the behaviorists in the mode of J. B. Watson, placed all their bets—at least as far as personality and its social correlates is concerned—with the impact of "environment." These theoreticians and their grass-roots adherents gave no consideration to the realm of heredity, often simply by not addressing the specific problem (e.g., mental retardation), or by seeking explanations which would derive *totally* from external causes. In some sense, psychoanalysis, in spite of its medical roots, further supported this line of reasoning. It did so by the preponderant emphasis on the impact of the childhood environment, and by its stress on the functional development of psychodynamic health or malfunction. All this, we may note parenthetically, in spite of the evidence of Freud's reluctance to his dying day to abandon his view of himself as a "neurologist," as a physician concerned with the specific functioning of the nervous system's physical reality.

We thus find a strangely mottled background of psychological thought, variously recast in the popular psyche, against which quite recent medical developments bearing on the potential control of genetic disease have made their appearance, including an often-confused acceptance of the presumably inevitable, the idealized and often misguided adulation of "hypothetically pure" and "uniformly healthy" artifactual populations, and the broad deemphasis of genetic cause, based on behavioristic or psychoanalytic doctrine. Together with simple lack of concern, the popular perspective on genetics is not encouraging.

In spite of this melancholy background of popularly held belief, significant change may be in the offing. A rather different contribution to popular thought relating to the role of genetics has recently emerged from fundamental advances in the basic theory of genetics itself. The discovery and unfolding knowledge of deoxyribonucleic acid (DNA), the chemical substance of genetic material as drawn to wide public attention by books such as *The Double Helix,* the notion that forms of life may be created or genetically manipulated, as by experimentation with recombinant DNA, and the present-day folklore relating to the possibilities of cloning (the exact reproduction of identical "persons" or other life forms) have brought renewed interest to, and confusion in, people's ideas about genetics and its uses. Genetics has been publicly "rediscovered," but once again scientific progress has come to be reflected in the minds of many by a melange of hope and anxiety, knowledge and misunderstanding.

At any rate, there are some indications that the interplay of medical advances and changes in personal and social values may be setting the stage for a realistic application of new medical technology for far-ranging human benefit. The medical technology alone, of course, cannot do the job unless the relevant people beyond the confines of the laboratory, are prepared to avail themselves of the new and rapidly evolving methods and their potential advantages.

Every human being carries between five and eight deleterious genes. These recessive genes have no negative consequences for the health of the carrier. However, being a carrier predisposes, under certain conditions, to development of genetic anomaly at certain risk levels in the carrier's fetus and child.

In this context, we must now turn to a more explicit description of certain major medical technologies that have become available in the past decade, in the purposeful control of genetic disease.

Two interrelated lines of development are observed: (a) an enhanced large-scale capability for screening—e.g., for determination of carrier status; (b) a vastly improved technology for the prenatal diagnosis of genetic anomalies, making possible early in fetal development consideration of alternative options, such as therapeutic abortion.

The availability of technology to provide often exceedingly reliable and valid information regarding the genetic health or dysfunction of the fetus makes it worthwhile to ascertain the potential parents' possible carrier status, as it may place a fetus at risk with respect to a specific genetic ailment. The issue of what is "worthwhile" may be addressed variously in statistical and in human terms–that is, in terms of the probabilities that being or not being a carrier may contribute to birth of defective offspring, and in terms of the benefits and/or liabilities (including possible stigma or anxiety) may affect persons identified as carriers.

Beyond these considerations, there is the issue of "knowing the worst"–the instance in which little or nothing can be done in specific medical terms, except perhaps by mild psychological or medical amelioration, to avoid ultimate tragedy. An instance of the latter appears, for example, under conditions of positive diagnosis of a fatal genetic illness *in utero* in which the parents choose not to, or cannot, avail themselves of therapeutic abortion; here, based on present medical knowledge and experience, the afflicted child, may well die at an early age, as in Tay-Sachs disease, though various temporary and supportive measures can be the only feasible interventions in the face of an unavoidable and tragic reality.

As relates to screening for carrier status and prenatal diagnosis, a basic schema of alternatives appears in Figure 1. Schematically, we may consider a chain of events which begins with the procedures identified in (a), above, denoted as screening, proceeding toward a variety of possible subsequent interventions–e.g. (b), denoted as prenatal diagnosis–and subsequent potential remedial actions.

The schema represents an oversimplification; the needed rectification is indeed our central concern. One must examine the *psychosocial context* in which the chain of events shown in Figure 1 unfolds. This context includes the attitude base motivating or demotivating compliance with suggested

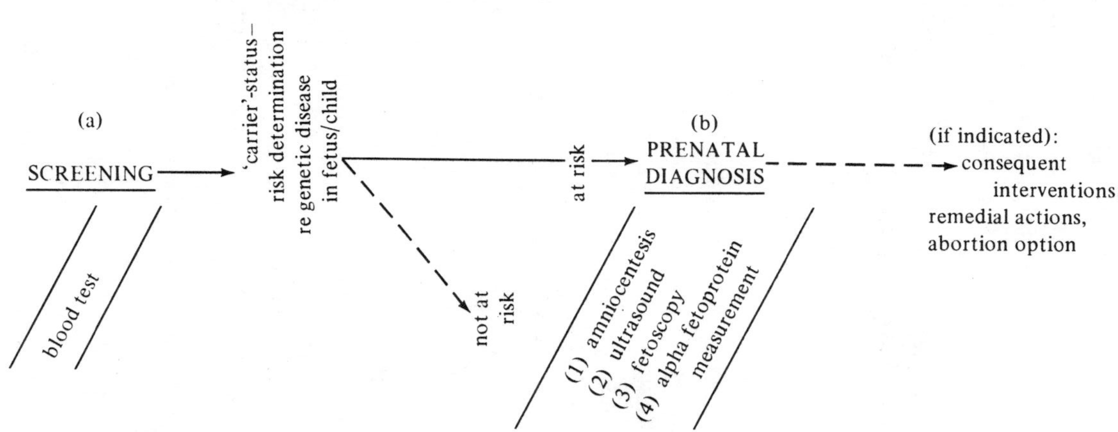

Figure 1: Schema: The Genetic Disease Control Process

screening for various genetic disorders, responses to the screening process itself and to its outcomes (i.e., carrier/non-carrier status), attitudes relating to mating and conception under various conditions of risk, views of prenatal diagnosis, response to positive or negative diagnostic findings, and eventually readiness to seek or to reject therapeutic abortion.

It is evident that a complex decision tree confronts the potential user of the evolving medical technologies, all imbedded in a complex framework of human response. This framework determines the ultimate personal and social effectiveness of the available technical means for genetic disease control.

This psychosocial context poses the emergent question, "How will this new medical/genetic technology be received by the public(s) that it is intended to benefit?" In light of correct and incorrect perceptions of the nature of genetics and its consequences, and within a pattern of changing social values, what is the probable impact of the varied linkage "screening→prenatal diagnosis→consequent intervention" on control of genetically caused illness, or well-being?

SCREENING

A substantial literature is available describing the concept and process of screening. For present purpose, a brief summary of typical screening efforts must suffice. In most instances, we shall report particularly on the California Tay-Sachs Disease Prevention Program as an example of screening for carrier status among adults. Other genetic screening programs seek to determine possible presence of a genetic illness in the person affected, especially among the newborn. Perhaps best known are screening programs for phenylketonuria (PKU) and sickle cell anemia.

Some definitions relating to these programs may be briefly cited (Committee for the Study of Inborn Errors of Metabolism, 1975: 9).

> Genetic screening may be defined as a search in a population for persons possessing certain genotypes that (1) are already associated with disease or predisposed to disease, (2) may lead to disease in their descendants, or (3) produce other variations not known to be associated with disease.
>
> The persons in the first category are identified so that medical management may be provided. The second group is discovered so that reproductive options may be discussed . . . the third category gathers information for research purposes—that is, for the study of the genetic constitutions of populations.

In formal terms, as the Committee for the Study of Inborn Errors of Metabolism notes in its discussion of screening methodology (1975: 231), the test (such as the blood test used for Tay-Sachs disease carrier screening and, with technical modification, for other predictable genetic anomalies) should be "reliable, repeatable, and accurate." It is precisely because tests are now available which in large measure meet these criteria that large-scale genetic disease screening has become a reality. For example, through June 1980, the California Tay-Sachs Disease Prevention Program screened 75,000 persons to determine their carrier status with reference to that disease. In seven countries worldwide, some 350,000 persons have to date (1980) voluntarily participated in Tay-Sachs disease screening programs.

A Tay-Sachs Disease Prevention Program Manual (Greenwald et al., 1975) describes in detail procedural and technical issues relating to programs of this nature. In its basic form (and we cite the Manual's introduction), the program is directed toward the following objectives:

> A. To inform individuals, parents, and prospective parents about Tay-Sachs disease and its prevention;
>
> B. To provide the community with a blood testing program for Tay-Sachs disease carrier detection;
>
> C. To provide counseling for carriers, relatives of carriers, 'at risk couples', pregnant women and their spouses, or any other individuals seeking information.

And, going beyond the blood test screening phase itself, addressing prenatal genetic diagnosis:

> D. To provide a means for intrauterine evaluation of pregnancies in couples at risk for this disorder in their offspring; and
>
> E. To follow up and evaluate the effectiveness, psychosocial impact and ill effect (if any) of such a program on individuals, families and/or the community at large.

Given the significantly higher incidence of Tay-Sachs disease in Jewish (principally Ashkenazic) populations, the Tay-Sachs disease screening program faces the further challenge of reaching, by voluntary compliance, a *sub*population within the minority population matrix—a subpopulation that is not conveniently identified and, indeed, whose very definition is subject to varied interpretation.

Among the most controversial of screening programs is that focused on sickle cell anemia and sickle cell trait. Sickle cell anemia is a blood disease in which red blood cells are found in an aberrant shape, resembling a sickle, rather than following the normal rounded form. Sickle cell trait constitutes a predisposition rather than a disease syndrome itself. Both conditions are found disportionately among Blacks, particularly among racial groups whose origins reach to Africa, with less clear-cut antecedents appearing at Mediterranean, Central American, and Indian Subcontinent location.[2]

While sickle cell conditions are predictable on a probability basis in accordance with hereditary principles, for procedural and interpretational reasons, U.S. screening programs in the sickle cell field have not proved to be unmitigated successes. Among complications has been confusion between sickle cell *disease* and sickle cell *trait,* as has been pointed out, among others, by Frost and Kaback (1973).

While, as indicated, emphasis here is placed on genetic screening of adults, as relates to their possible status as carriers of genetic disease, one also needs to note widespread testing of newborns for disorders such as PKU, a genetically determined cause of mental retardation. Normally, a procedure known as the Guthrie test is employed. In this type of screening, a remedy for the disease state is sought by implementation of a specific corrective dietary regimen.

A vast literature has been available relating to PKU since its discovery as description by Folling in 1934. Mass newborn screening for this disease dates to the early 1960s. A useful nontechnical summary of the nature of PKU and other testing procedures is provided in Levy (1973).

As often is the case, a specific discovery, while accurately described, is not widely recognized and impactful until some other discovery makes its appearance and establishes synergy with the former. Mendelian heredity, with its experimental and conceptual identification of the gene, gained substantial recognition only many years following its initial statement, when other discoveries came onto the scene. As Scriver et al. (1978) noted, a conjunction of this kind occurred when the physician A. E. Garrod drew attention to the relevance of Mendelian concepts for an understanding of human health and disease. From these beginnings in 1902 developed the increasing recognition, joined by other discoveries in biology, chemistry, medicine, and other fields, that specific interventions, such as screening for genetic ailments and prenatal genetic diagnosis can become practical and widely applicable procedures.

PRENATAL GENETIC DIAGNOSIS

While genetic screening addresses issues such as carrier status in adults (as in Tay-Sachs disease), or the presence of a genetic defined inborn metabolic error in the newborn (as in PKU), prenatal diagnosis provides indications, by examination of the intrauterine environment, of the fetus's possible genetic affliction or health. Such intrauterine diagnostic procedures can serve to predict, in some instances, the organism's eventual biological fate—for instance, death at an early age.

Ream (1978) notes that "the effects of modern instrumentation on medical practice have been profound. There is little doubt that they have substantially altered earlier limitations on the quality of care" (1978: 963). Arguably, there is hardly a field of medical practice in which this statement is more clearly valid than in clinical genetics. An impressive array of procedures has been developed in the past decade to make highly reliable diagnosis of genetic disorders *in utero* a reality.

Foremost among the technologies used in prenatal diagnosis of genetic diseases or defects are (1) amniocentesis, (2) ultrasound (pulse-echo sonography), (3) fetoscopy/placental aspiration, and (4) alpha fetoprotein measurement.

Amniocentesis

Amniocentesis—a simple outpatient procedure—involves sampling the amniotic fluid at about 16 weeks' gestation to ascertain the sex, chromosomal, and certain genetically defined biochemical parameters of the fetus in pregnancies. The procedure is carried out particularly if there is a significant risk of genetic disease.

As generally performed, the technique involves transabdominal aspiration of a sample of fluid of the amniotic sac under local anaesthesia. Upon withdrawal of fluid from the amniotic sac, one derives from the fluid sampled a culture of cells, to permit analysis of the chromosomes (karyotyping) and enzymes (biochemical test).

At a high level of assurance, amniocentesis as it has evolved since 1968 has proved to be a safe procedure, with a very low proportion of complications, such as vaginal bleeding or amniotic fluid leakage. Nor does it appear to induce any disproportionate likelihood of spontaneous miscarriage or fetal death in experienced hands. Rather, as noted in the task Force report of the National Institutes of Health (1979), the establishment of a clear indication of likelihood that a genetically diseased child will be born has created new reproductive options for the families affected. The report notes that, while amniocentesis has been widely applied for fetal cytogenetic study, including assessment of Down's Syndrome, the method is "not a procedure which is deemed applicable to all pregnancies, nor is it a means for assuring a family of the complete absence of any hereditary disease or congenital defect in the fetus." However, it is evident that amniocentesis has come to be a method of prime importance in the prediction of certain genetic disorders and indeed, as Omenn (1978) has noted, this procedure has "revolutionized the practice of clinical genetics and genetic counseling."

Ultrasound

In industrial settings, there has been increased emphasis on "nondestructive testing" of products whose interior parts are difficult to access. Sparked by the same technological roots, including radar and sonar, it has now become possible to "reach" deep inside the human organism without the necessity of direct physical intervention, such as surgical incision or needle insertion. Building on common foundations, as evolved in the 1940s and 1950s, ultra-high-frequency sound, supplemented by sophisticated imaging techniques, provides the skilled observer with pictures *inside* otherwise difficult-to-reach portions of the human body, in graphic and meaningful form.

At this stage of development, the procedure of using ultrasonic waves in medical diagnosis by use of sound waves with frequencies of millions of hertz (cycles per second) and by ancillary displays and scanning systems is costly. However, as noted in National Institutes of Health (1979: 86-87):

> Ultrasonography has and will have an impact on the care of pregnant women during the first 20 weeks of gestation.
>
> 1) It increases the accuracy of the diagnosis of fetal disease.
>
> 2) It decreases or eliminates the need for x-ray exposure during pregnancy.
>
> 3) It can allay anxiety in concerned parents because [for] the vast majority . . . evaluations indicate no detectable abnormality in the fetus.
>
> The full potential of ultrasonography probably has not yet been comprehended . . . it may take 20 more years of development but it seems probable that a high percentage of . . . children with physical malformations will be recognized before the twentieth week of gestation.

A substantial overview of the field's current status, of interest to the general reader, appears in Devey and Wells (1978).

Fetoscopy

For fetal diseases that cannot be readily diagnosed by means of amniocentesis, alternative approaches have been developed, making possible direct viewing of the fetus, or direct entry into the uterus to obtain blood or tissue samples.

With origins in the 1950s, and further development in the two decades following, fetal visualization and blood sampling have demonstrated high levels of diagnostic accuracy. However, current uses of fetoscopy are limited, though future and more energetic experimentation with the method is anticipated, particularly in connection with a genetic disease known as "beta thalassemia," and with certain cases of sickle cell anemia.

Alpha Fetoprotein

Also associated with the use of amniotic fluid as source, alpha fetoprotein procedures are employed to detect neural tube defects *in utero*, primarily in women with previously affected pregnancies. These are serious, often fatal, anomalies resulting from failure of closure of the skull or spinal column in the fetus. The factor AFD is found to be greatly elevated in the amniotic fluid when such anomalies are present. Such defects have significant economic consequences and costs, associated with genetically caused paralysis, congenital heart defects, and mental retardation (National Institutes of Health, 1979).

SUMMARY

We have briefly described above, citing both medical and popular sources, several well-developed and rapidly developing approaches to the prenatal diagnosis of genetic disorders. Of these, amniocentesis is perhaps the procedure of widest current utility, with increasing promise held for ultrasound, particularly as this noninvasive procedure comes to be technically simplified and less costly. At any rate, the rapid and still-accelerating rise of these medical technologies, providing increased hope and opportunity for the control of genetic disease, depends in their ultimate impact on a wide range of psychosocial factors.

NOTES

1. A detailed statement of Tay-Sachs disease screening and prevention appears in Kaback (1977b). Other citations relating to Tay-Sachs disease screening programs appear in the References at the back of this book.

2. A useful bibliographic survey on sickle cell anemia can be found in Davis (1975).

Chapter 2

THE PSYCHOSOCIAL STRUCTURE OF GENETIC DISEASE CONTROL

The rapid development of medical technologies suitable for prediction and diagnosis of genetic disease points to the need for a psychosocial theoretic framework to explain and further develop effective utilization of this technology. With particular attention to the Tay-Sachs disease prototype, this chapter provides a working model for such a framework.

As noted, empirical data for a specified Jewish subpopulation are reported. These data can be interpreted within the scope of systematic Jewish population research, such as the National Jewish Population Study.[1] They cannot be generalized uncritically, however, to other ethnic populations, especially in view of the Jewish population's high educational and socioeconomic level, which significantly exceeds that of many other groups. Still, an examination of those "cells" in the statistical tables that include, for instance, relatively less educated or less affluent Jewish respondents provides some hypotheses as starting points for theory formation and eventual empirical investigation in contrasting differentiated ethnic populations. The latter focus is of special relevance as certain genetic disorders appear with disproportionately higher frequency in identifiable "ancestry" groups. For instance, while Tay-Sachs disease is more clearly linked to Ashkenazic Jewish origin, beta thalassemia is relatively more prevalent in populations of Mediterranean origin, while sickle cell anemia appears predominantly in Blacks in the United States. More than 100 other hereditary disorders have been shown to occur predominantly in defined ethnic, racial, or national groups. For the most part, this reflects long-standing preferential inbreeding within these populations.

In proposing a theoretic frame for the psychosocial study of genetic disease control, it is recognized that, particularly in the 1970s, against a previously barren background, significant empirical work was commenced on psychosocial topics relating to genetics and genetic disease control.

In 1972, in the *Annual Review of Medicine,* Barton Childs observed that "there is no doubt that the behavioral sciences have flourished without benefit of genetic theory or thought." As he considers a variety of factors (e.g., intelligence, mental abilities, personality, psychotic behavior) that might be illuminated by recourse to genetic principles, Childs also examines (through 1970) the extent to which selected psychologically oriented journals had published papers on genetic subjects. It becomes quickly evident that certain "mainstream" psychological publications, such as the *Psychological Review,* the *Psychological Bulletin,* and the *American Psychologist,* had virtually ignored topics relating to genetics. In turn, it may be noted that the extent to which the medical profession and geneticists had made use of psychological and sociological modes of inquiry left much to be desired. Subsequently, certain subfields, including Tay-Sachs disease prevention and patient compliance with medical recommendation, have come to make effective use of psychosocial and behavioral science research and concepts.

The available researches are based on various assumptions regarding relationships between what goes on in the genetic/medical realm (e.g., the nature of genetic disease, new technologies available for control of genetic disease, and so on), and how people act, or are likely to act, under various circumstances (e.g., how what people know about genetic disease affects their behavior, when and how they are motivated to take part in a genetic screening program). Some studies formulate "mini-models," indicating the character of the expected relationships and report data collected within these frameworks. The issue becomes one of identifying conceptually and testing empirically the relationships between one set of conditions or events occurring in genetic medicine and medical technology, on one hand, and on the other in the thoughts, feelings, and actions of the human beings actually or potentially affected. Among these there are, of course, the intended consumers of the available new services, such as blood test screening and amniocentesis, as well as the physicians, nurses, and others who deliver the specified services. Evolving research topics such as health behavior, psychosocial aspects of health service delivery, and sociological research examining the dynamics of the associated institutional contexts such as hospitals, clinics, private physician offices, and HMOs illustrate inquiry in this area.[2]

In this context, we may for present purposes distinguish a *psychosocial structure,* including attitudes and behavior (as well as feelings and underlying motivations) of relevant publics, and a *genetic/medical structure.* Each affects the other, and thus each is both independent and dependent variable.

As we consider the patterns of these structures, some caveats immediately are in order: the structures themselves, though we may hold them constant for suitable analytic purposes, are continually in *process,* some elements relatively stable, others changing at a rapid pace. For instance, given aspects of genetic disease etiology may, for some considerable periods of time, be regarded as correct, with treatment and clinical management following specified normative procedures with considerable clarity. However, the exceptional growth in fields such as development of medical technology for diagnosis and prediction of genetic illness points to new and different sociomedical approaches. At one time, nothing could be done about certain conditions; with medical technological innovation, control increasingly becomes possible. However, such control cannot occur unless the relevant *psychosocial processes* are understood and mobilized in utilization of this technology, especially if use is voluntary. It is evident that a technology, no matter how sophisticated and potentially useful, will not have much impact in the real world unless the constellation of human processes is in place to put the technology into action.

As a further caveat, it is necessary to recall that any framework is useful only to the extent that it provides a basis for conceptual articulation and systematic study of a phenomenon of *interest.* The very choice of research interest stems from a value system, and this value system continues central as the framework is defined and applied. With these cautions in mind, we proceed with development of a working schema.

We consider three major *concept categories* which, in combination, determine the effective control of genetic disease:

(a) *genetic disease control technologies* (e.g., blood test screening; prenatal diagnosis, such as amniocentesis)
(b) *communication flow,* transmitting information regarding genetic disease and its control (e.g., mass media, scientific journals, physicians, friends, and the like)
(c) *personal readiness,* by people in relevant publics, to make use of available genetic disease control technology, (e.g., specified demographic characteristics, education levels, value patterns)

If certain optimal conditions occur in each of these concept categories, we may predict high probabilities of genetic disease control. Of course, the *specific* conditions to be met in each category need to be specified, and they vary for different publics and for different genetic diseases. In turn, genetic disease control technologies and information flow patterns are differentially suitable for different publics and diseases, to motivate desired behavior–i.e., utilization of the technologies, in the cause of genetic disease control.

GENETIC DISEASE CONTROL TECHNOLOGIES

In Chapter 1, we provided a brief description of new genetic disease control technologies, generally developed in the 1960s and 1970s, providing heightened capabilities for the prenatal identification of genetic defects. In addition to these technologies, we consider in the present rubric two other methods, one of current medical standing, the other of ancient origin.

The former is the procedure of *blood test screening,* making it possible to specify, at very high probability levels, whether a given person is or is not a carrier of a specific genetic disease gene or in some instances directly involved with the genetic disease. Here, we consider, by way of example, the process of Tay-Sachs disease screening.

A succinct description of Tay-Sachs disease, including a brief statement of the disease's history and mythology appears in Kaback (1977a: 1-5). In children with the disease–who die by the time they are three or fours years old following an apparently normal birth–a complete deficiency is found in the nervous system of an enzyme catalyst known as hexosaminidase A or "hex A." This catalyst is required to degrade a normally produced cellular lipid. The hex A deficiency causes progressive accumulation of this fatty material within brain cells, leading to gradual degeneration of the child's nervous system. If two carriers of the Tay-Sachs disease gene conceive, there is a chance of *one in four* that a child afflicted with the disease, with a prospect of certain death, will be born. The blood test screening procedure, therefore, can serve to identify carriers, making possible, with the aid of genetic counseling, appropriate family planning. Of course, in instances in which birth of a Tay-Sachs child is considered a possibility, the genetic disease control technologies in Chapter 1 may be employed to determine whether any given specific fetus is diseased, providing the parents with information that establishes an informed basis for choice of therapeutic abortion if desired, or for continuation of pregnancy.

COMMUNICATION FLOW

We consider the following process: a set of genetic disease control technologies is available (A). These include, most significantly, blood test screening and prenatal diagnosis, such as amniocentesis. Information flow (B), regarding the availability and potential uses relating to these methods as means for the control of genetic disease, needs to be directed toward the potential user or "consumer" of this technology (C). In the rubric of communication flow, we include both typical "media" sources and education and social structures by which information is mediated. This subject will be developed at length in the subsequent chapter.

PERSONAL READINESS

However, even the most substantial flow of communication has no effect unless it is received by people in the appropriate public at a suitable level of personal readiness, for understanding and action. This readiness presumably must prevail in a population base of adequate size, to warrant promulgation of the specified information.

Figure 2 provides a conceptual representation of the process indicated. The figure shows the major steps, variables, and outcomes of the process, including the interaction between the availability of genetic disease control technology, communication flow, and personal readiness for response to this technology. Specific alternatives include awareness of and participation in screening (compliance with recommend-

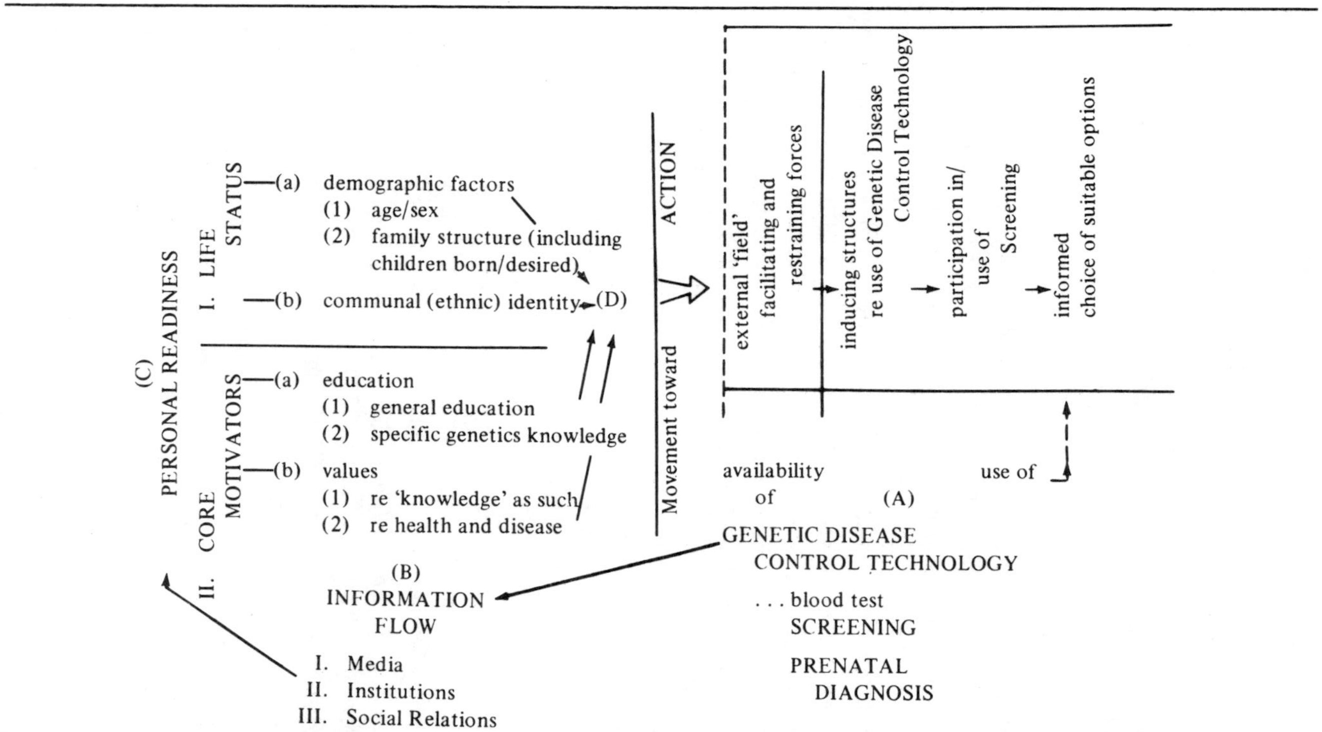

Figure 2: Schema: Alternatives, Screening for Carrier Status and Prenatal Diagnosis

ed procedure) in the California Tay-Sachs Disease Prevention Program. Recognition of carrier/noncarrier status of the specific genetic disease, conception/nonconception, and as indicated use of genetic disease control technology in instances of risk are additional stages in the process. Finally, there remains the issue of selective abortion, if prenatal diagnosis determines the presence of genetic disease in the fetus.

The issue of personal readiness, leading to genetic disease control, is complex. We need to examine several *strategic* variables from among the many that may be identified.

In prior research, Rosenstock (1966) elaborated a "health belief model." This model is concerned with a specification of factors associated with actions of presumably healthy persons toward disease prevention or detection of disease prior to appearance of symptoms. It focuses on psychological states of readiness to take such action and on perceptions by the subjectively healthy individual that a specified action would effectively reduce a particular disease threat. Perceived susceptibility and perceived seriousness are seen as significant determinants for initiation of preventive action.

While the schema here proposed concurs with Rosenstock that readiness is a critical variable, it holds that such *readiness may relate more saliently to general knowledge and ideologic variables, rather than primarily to variables that are clearly related to perception of a particular disease and its possible impact on the individual.* The work of McQueen (n.d.) and that of Goldstein et al. (1977) suggests that, at least in prototypic research on Tay-Sachs disease, such factors as *Jewish identity* and *general knowledge,* as well as various demographic factors, rather than perceived susceptibility or perceived severity of the disease are critical forces motivating participation in a screening program.

The issue, however, may not be the inherent "correctness" of the health belief model as such. Rather, it may be necessary to view this model in the context of a *systems* framework, noting the relative position of its prime variables vis-à-vis other variables in a more dynamic context of inputs and outputs, and mutual interactions. As focused on personal readiness, we propose the following complexly interactive subsets of variables within which those specified in the health belief model are contextually embedded:

I. Life Status
  A. Demographic factors
    (1) age/sex
    (2) family structure (including children born/desired)
  B. Communal (ethnic) identity
II. Core Motivators
  A. Education
    (1) general education
    (2) specific genetics knowledge
  B. Values
    (1) re 'knowledge' as such
    (2) re health and disease
(a)   (a) re susceptibility to specific disease
      (b) re severity of specific disease
      (c) re stigma of association with specific disease or carrier status
      (d) re the nature of life (conception, abortion, life duration)

We hypothesize that, for particular genetic disease, the variables noted, *including but not confined to those of the health belief model,* can account significantly for personal readiness conditions required for the effective use of newly available genetic disease control technology. The following examples may be considered:

In Tay-Sachs disease prevention, personal readiness for the relevant genetic disease prevention technology will be high if the following conditions are met:

I. Life Status
  A. Demographic factors
    (1) age/sex: in reproductive-age range (some possible sex difference, favoring females)

(2) family structure: some additional children desired/expected; number of children born to couple–0 or 1

B. Communal (ethnic) identity: Jewish ethnic identity affirmed (Tay-Sachs regarded as "Jewish disease")

II. Core Motivators

A. Education

(1) general education level high

(2) genetics knowledge level high

B. Values

(1) re knowledge as such: knowledge of health conditions viewed as valuable in its own right

(2) re health and disease:

(a), (b) issues of susceptibility and severity of marginal significance;

(c) no special concern with stigma attached to possible carrier status,

(d) re the nature of life: life valued more highly for those "already born" rather than for unborn fetus; "Life is for the living" (outside womb) as important value statement.

In cystic fibrosis control, in general Anglo populations:

I. Life Status

A. Demographic factors

(1) age/sex: in reproductive-age range, taking account of subcultural variations–e.g., possible younger age of mother at birth of first child

(2) family structure: some additional children desired/expected; 0, 1, or possibly 2 children born (though latter figure least predisposing toward personal readiness)

B. Communal (ethnic) identity not particularly well crystalized, may be heightened by information inflow

II. Core Motivators

A. Education

(1) general education level medium to high for optimum response

(2) genetics knowledge level medium to high for optimum response; as necessary raised by specific information inflow, campaigns

B. Values
   (1) whether or not knowledge for its own sake is initially valued, heightening of this value by special information inflow, campaigns
   (2) health
      (a) perceived susceptibility to disease
      (b) perceived severity of disease quite high
      (c) perceived stigma of being disease carrier, while fairly high, to be lowered by special information inflow, campaigns
      (d) value of life, as may favor unborn fetus over living parents (esp. mother) as possible barrier to utilization of elective abortion, and thus possible obstacle to use of certain genetic disease control technologies—e.g., "What God wills must be endured."

NOTES

1. For instance, see Chenkin (1972); Massarik and Chenkin (1973); Massarik (1974-1975, 1980).

2. Some relevant citations are: Becker and Maiman (1975); Becker et al. (1974); Marston (1970); Beck et al. (1974). Focusing on Tay-Sachs disease as a useful prototypic case in point, see Childs et al. (1976, n.d.); Seashore et al. (1976); Becker et al. (1975); Margolin (1978); McQueen (n.d.); and Goldstein et al. (1977).

Chapter 3

GENETICS KNOWLEDGE:

IMPLICATIONS FOR DISEASE MANAGEMENT

In the development of voluntary programs directed to the prevention and control of genetic disease, the public's level of knowledge of such disease assumes special importance. In the framework of the schema discussed in Chapter 2, it is likely that some minimum "threshold" of valid genetics knowledge, together with appropriate demographic and motivational conditions, is essential to assure voluntary participation in programs such as community-based Tay-Sachs heterozygote screening. Further, such knowledge may be prerequisite to effective utilization of genetic counseling and relevant in shaping attitudes regarding the manageability of genetic disease. In this context, answers to the following questions assume particular significance:

(1) How is the knowledge level of genetic principles and concepts to be defined and conveniently measured?
(2) What is the prevailing genetics knowledge level in specified populations?
(3) What education levels and other demographic characteristics are differentially associated with various genetics knowledge levels?

(4) What is the relationship between genetics knowledge level and views of manageability of genetic disease?

At this time, the medical profession and general public lack a practical measure of genetics knowledge. For this purpose, the *Genetics Knowledge Index* (GKI) is constructed and employed, initially on a trial basis. The GKI is defined by equally weighted responses to 22 questions on principles of genetics and genetic illness, as selected from a mail survey questionnaire administered, with nonresponse follow-up, in a study of Jewish persons (n = 1018) in Los Angeles in 1974. The method, described further in Chapter 8, is based on the following assumptions: (a) Because of the disproportionately high frequency of appearance of the Tay-Sachs disease (TSD) gene in Jewish populations (especially among those of Ashkenazic origin), chosen samples focus particularly on Jewish populations; (b) because TSD control is especially germane for persons planning to have children (for the first time or in addition to those already born), emphasis is placed on those of reproductive age; (c) because of its high education level and typical responsiveness to community programs, the Jewish population constitutes a useful "test case" in genetic disease control. If voluntary participation in a genetic disease screening program were to fail here, even less success could be expected under less favorable conditions, in other population groups. The sample was so designed as to provide maximum representation of the target population, per above, rather than seeking a conventional cross section of the general population or of the Jewish population as such. Likewise, response selectivity, as demonstrated by the telephone follow-up to those initially not replying, favors desired access to the target population—high proportions of Jewish respondents of reproductive age were reached, representing the desired prototype (for test items, see Appendix A).

While the GKI is viewed as a convenient heuristic device for practical assessment of genetics knowledge, particularly in survey and field-interview research, it is here presented as starting point for further field experimentation and reliability and validity determination.

Five genetics knowledge categories, referred to as GKI levels, are specified:

number of correct responses: 0-10, level I (lowest)

number of correct responses: 11-13, level II

number of correct responses: 14-16, level III

number of correct responses: 17-19, level IV

number of correct responses: 20-22, level V (highest)

Supplemental data show the professionally diversified character of the subpopulation; there is no disproportionate incidence of persons whose occupations require specialized knowledge of genetics, such as physicians or related medical specialists.

With the exception of the 30-39 age range, superior GKI attainment (levels IV and V) appears with similar frequency for males and for females: in the 29-29 age group, 49 versus 54%; in the 40-49 age group, 43 versus 40%. However, in the mid-years, ages 30-39, males with their considerable education accomplishment, reveal high GKI scores in somewhat greater proportion than females; 51 versus about 40% (diff. sig. at .01).

In interpreting the relationship between demographic characteristics such as age/sex and education and GKI, we must consider that both emotional or psychological "distance" from conception and childbearing, as well as education-linked factual information, necessarily interact in determining resultant genetics knowledge levels.

A clear-cut relationship exists between general education level and genetic knowledge; the relationship is as expected: Low general education level is significantly associated with low GKI; high general education level is markedly linked to high GKI scores. In a sense, this association serves as a measure of validation.

TABLE 1 GKI Distribution: Age/Sex and Genetics Knowledge

	Males			*Females*			*Totals*		
GKI	*20-29*	*30-39*	*40-49*	*20-29*	*30-39*	*40-49*	*Males*	*Females*	*Males & Females*
I	(2) 2.5	(14) 6.0	(18) 18.8	(2) 1.3	(17) 7.3	(6) 11.5	(34) 8.4	(25) 5.7	(59) 7.0
II	(9) 11.4	(24) 10.3	(9) 9.4	(15) 9.6	(42) 18.1	(8) 15.4	(42) 10.3	(65) 14.8	(107) 12.6
III	(29) 36.7	(75) 32.3	(28) 29.2	(55) 35.3	(81) 34.9	(17) 32.7	(132) 32.4	(153) 34.8	(285) 33.6
IV	(30) 38.0	(96) 41.4	(38) 39.6	(71) 45.5	(88) 37.9	(19) 36.5	(164) 40.3	(178) 40.5	(342) 40.4
V	(9) 11.4	(23) 9.9	(3) 3.1	(13) 8.3	(4) 1.7	(2) 3.8	(35) 8.6	(19) 4.3	(54) 6.4
Total n %	(79) 100.0	(232) 100.0	(3) 100.0	(156) 100.0	(232) 100.0	(52) 100.0	(407) 100.0	(440) 100.0	(847) 100.0
Summary:									
% I + II	(11) 13.9	(38) 16.3	(27) 28.2	(17) 10.9	(59) 25.4	(14) 26.9	(76) 18.7	(90) 20.5	(166) 19.6
% IV + V	(39) 49.4	(119) 51.3	(41) 42.7	(84) 53.8	(92) 39.6	(21) 40.3	(199) 48.9	(197) 44.8	(396) 46.8
	X^2, sig.*			X^2, sig.**					

NOTE: Figures for age/sex cohorts with highest GKI levels (40% or over at specified levels) underlined.

NOTES TO INTERPRETATION OF TABLES: In statistical data reported, unless otherwise indicated, the following conventions will be maintained:

a. As appropriate, levels of statistical significance typically for chi-square, with adjustment for continuity as may be required, will be denoted with ** at significance level of .01 or better, and with * for significance level attaining .05. Lack of significance, within the confines of this statistical procedure, will be denoted n.s. Beyond observation of statistical significance and noting empirical focus on a fairly homogeneous subpopulation, findings suggestive of certain relationships and useful in generation of further hypotheses also require purposeful consideration.

b. Percentage totals may not add to exactly 100 due to rounding. However, they will be shown as 100 if differences due to rounding only affect the numerical total.

c. Numbers of cases are shown in parentheses; corresponding percentages are rounded to one decimal.

d. While the initial mail questionnaire sample reached 1018 cases, with the study's principal emphasis on persons of reproductive age, most tables exclude other age cohorts. In some instances, on topics of particularly Jewish relevance, non-Jewish persons are not included. Therefore the n's in most tables are smaller than the actual response n.

e. Number of cases and procedures in carrier follow-up study phase are reported in Chapter 7.

As is shown in Table 2 (in this instance extending the age distribution beyond the reproductive range), among those with less than 12 years of schooling, some 87% score in the lowest or low GKI range (levels I and II), while those with advanced university education (professional degree) include about 88% who score in the middle, high, or highest GKI

TABLE 2 Education Level and GKI

		less than 12 yrs.	*high school*	*some college*	*college grad.*	*profess. degree*	*Total*
GKI	I	65.2 (15)	26.8 (22)	9.3 (24)	7.2 (23)	4.0 (10)	10.1 (94)
	II	21.7 (5)	28.0 (23)	23.6 (61)	9.7 (31)	7.7 (19)	14.9 (139)
	III	4.3 (1)	29.3 (24)	32.4 (84)	36.3 (116)	32.0 (79)	32.6 (304)
	IV	8.7 (2)	15.9 (13)	32.4 (84)	41.6 (133)	43.3 (107)	36.5 (340)[a]
	V	0 (0)	0 (0)	2.3 (6)	5.3 (17)	13.0 (32)	5.9 (55)
	Total	100.0 (23)	100.0 (82)	100.0 (259)	100.0 (320)	100.0 (247)	100.0 (931)
		2.6	8.8	27.8	34.3	26.5	100.0

NOTE: Table includes unmarried respondents.

a. one case with education level not indicated eliminated from computation

χ^2**

categories, (levels III, IV, and V). There are very few "surprises"—only 8.7% (or two cases) among those with less than high school education score high on the GKI, while conversely only 11.7% (or 29 cases) among those with professional degrees score in the relatively low GKI range. Further, the education level of the Jewish population studied is extremely high: Nearly 61% are college graduates or have been awarded professional degrees beyond the baccalaureate level. An additional 28% have had some college education.

One cannot assume that high general education necessarily leads directly to high GKI. More likely, such general education, while already including some knowledge of genetics, additionally sets the stage for further learning and openness to supplemental information on genetics as may be made available subsequently, by media and by special educational campaigns.

ELEVEN GENETIC KNOWLEDGE ISSUES

To provide a further in-depth view of genetics knowledge, we now focus on eleven survey questions, included in the

GKI. A subset of GKI questions is analyzed below, in view of their substantive significance and to avoid redundancy. Among those not shown here are various questions relating to perception of the genetic character of various diseases (see Chapter 5).

The questions considered here are paraphrased and briefly designated as follows:

Question	*Paraphrase*	*Brief Designation*	*Correct Answer*
II-2	"terms inherited and congenital synonymous"	"inherited/congenital"	(no)
II-3	"definition of a gene"	"gene definition"	(a chemical unit of hereditary information)
II-4A	"genetic diseases can skip generations"	"skip generations"	(yes)
II-4B	"genetic diseases can be inherited from *one* parent only"	"inheritance/one parent"	(yes)
II-4C	"certain genetic diseases require inheritance from both parents"	"inheritance/both parents"	(yes)
II-4D	"most genetic diseases due to new mutations in affected person"	"mutations"	(no)
II-9	"genetic disease carrier can be healthy"	"carrier healthy"	(sometimes or always)
II-10	"genetic disease occurrence almost exclusively in males"	"in males only"	(yes)
II-11	"genetic disease occurrence almost exclusively in females"	"in females only"	(yes)
II-12	"genetic disease occurence almost exclusively in selected populations"	"in select populations"	(yes)
II-14	"meaning of 25% risk"	"25% risk"	(each pregnancy, 1 chance in 4 affected)

TABLE 3 Correct Response Summary Principles of Genetics and Inheritance

		Males	*Females*	*Total*	*Range*[a]	*Trend*[b]	χ^{2c}
II-2	"inherited/ congenital"	(251) 61.7	(313) 71.1	(564) 66.6	60.4-82.7	none	**
II-3	"gene definition"	(269) 66.1	(246) 55.9	(515) 60.8	40.4-68.4	older less informed	**
II-4A	"skip generations"	(372) 91.4	(416) 94.5	(788) 93.0	90.6-95.5	older less informed	n.s.
II-4B	"inheritance/ one parent"	(374) 91.9	(404) 91.8	(778) 91.9	88.5-94.2	none	n.s.
II-4C	"inheritance/ both parents"	(274) 67.3	(306) 69.5	(580) 68.5	52.1-77.6	older less informed	n.s.
II-4D	"mutations"	(247) 60.7	(214) 48.6	(461) 54.4	44.2-63.4	none	***
II-9	"carrier healthy"	(392) 96.3	(428) 97.3	(820) 96.8	94.2-97.9	older less informed	n.s.
II-10	"in males only"	(300) 73.7	(306) 69.5	(606) 71.5	65.4-76.9	older less informed	n.s.
II-11	"in females only"	(260) 63.9	(240) 54.5	(500) 59.0	50.0-64.6	none	*
II-12	"in subpopulations"	(377) 92.6	(404) 91.8	(781) 92.2	84.4-100.0	older less informed	n.s.
II-14	"25% risk"	(392) 96.3	(424) 96.4	(816) 96.3	90.6-100.0	none	n.s.

a. percentage range for age/sex subcategories

b. *Trend* "older less informed" indicates consistent pattern, regardless of numerical magnitude of differences, in which, both for males and females, the percentage of correct responses decreases as age (per data for age categories 20-29, 30-39, 40-49) increases.

c. χ^2; (df = 1); differences between males and females designated as follows:

$\leqslant p$ .001 ***
$\leqslant p$ .005 **
$\leqslant p$ .01 *
not significant n.s.

In considering knowledge levels relating to genetic principles and human genetic disease, as indicated by the above questions, three alternative response patterns may be distinquished: (1) The respondent may possess the correct information; (2) the respondent may be misinformed; or (3) the respondent may be uncertain or may fail to provide an answer. Data reflecting these patterns are shown in Tables 3 and 4.

As indicated in Table 3, for both males and females the percentages of correct response appear substantial. Detailed findings concerning correct response taking account of age and sex difference appear in Table 4.

TABLE 4 Correct Responses: Detail Principles of Genetics and Inheritance

		Males 20-29	Males 30-39	Males 40-49	Females 20-29	Females 30-39	Females 40-49	Total Males & Females	χ^{2a} 'm'	'f'
II-2	"inherited/ congenital"	(52) 65.8	(141) 60.8	(58) 60.4	(108) 69.2	(162) 69.8	(43) 82.7	(564) 66.6	n.s.	n.s.
II-3	"gene definition"	(54) 68.4	(158) 68.1	(57) 59.4	(106) 67.9	(119) 51.3	(21) 40.4	(515) 60.8	n.s.	***
II-4A	"skip generations"	(73) 92.4	(212) 91.4	(87) 90.6	(149) 95.5	(219) 94.4	(48) 92.3	(788) 93.0	n.s.	n.s.
II-4B	"inheritance/ one parent"	(74) 93.7	(215) 92.7	(85) 88.5	(142) 91.0	(213) 91.8	(49) 94.2	(788) 91.9	n.s.	n.s.
II-4C	"inheritance/ both parents"	(58) 73.4	(166) 71.6	(50) 52.1	(121) 77.6	(153) 65.9	(32) 61.5	(580) 68.5	***	n.s.
II-4D	"mutations"	(46) 58.2	(147) 63.4	(54) 56.3	(85) 54.5	(106) 45.7	(23) 44.2	(461) 54.4	n.s.	n.s.
II-9	"carrier healthy"	(77) 97.5	(223) 96.1	(92) 95.8	(153) 97.9	(226) 97.4	(49) 94.2	(820) 96.8	n.s.	n.s.
II-10	"in males only"	(60) 75.9	(170) 73.3	(70) 72.9	(120) 76.9	(152) 65.5	(34) 65.4	(606) 71.5	n.s.	*
II-11	"in females only"	(51) 64.6	(148) 63.8	(61) 63.5	(95) 60.9	(116) 50.0	(29) 55.8	(500) 59.0	n.s.	n.s.
II-12	"in subpopulations"	(79) 100.0	(217) 93.5	(81) 84.4	(146) 93.6	(214) 92.2	(44) 84.6	(781) 92.2	***	n.s.
II-14	"25% risk"	(79) 100.0	(226) 97.4	(87) 90.6	(156) 100.0	(218) 94.0	(50) 96.2	(816) 96.3	n.s.	n.s.

NOTE: Correct response percentages 80.0% and above underlined.

a. for age differences: 'm' among males, 'f' among females: .001 ***; .05 *; not significant n.s.

Relatively least understood are the principles associated with questions II-4D, "mutations"; II-11, genetic disease occurrence "in females only"; II-3, "gene definition"; and II-2, the difference between the concepts "inherited/congenital."

While question wording and question structure need to be considered in interpretation of the results, it would appear that certain technical matter (re concepts as mutations, gene, and the like) and principles of inheritance in disorders affecting females are least understood.

Very high understanding prevails with reference to the following: II-9, that a carrier still may be healthy; II-4A, that genetic disease can skip generations; II-12, that it can occur particularly in certain subpopulations; II-14, the meaning of 25% risk in disease occurrence of future offspring; and II-4B, that genetic disease inheritance may be due to one parent. In these instances, correct responses exceed 90%.

While the above refers exclusively to distribution of *correct* responses, one wonders whether failure to provide correct replies is due mainly to actual *mis*information, or whether it primarily reflects uncertainty or doubt on the part of the respondent. A schematic representation of the response patterns bearing on misinformation and uncertainty appears in Table 5. *Misinformation* percentages in the 10.0 to 19.9% range are designated as x, those at 20.0% and above as xx. Similarly, *uncertainty* percentages (the total of "don't know" responses and "no answers") in the 10.0 to 19.9% range are designated?, while those at 20.0% and up are shown as ??. The following results appear (also see Table 5):

(1) The concept "gene" is rather widely misunderstood, in the sense that misinformation is reported. Also, misinformation surfaces quite frequently when respondents are asked to distinguish "inherited" and "congenital" disorders (II-3, II-2).

(2) While the level of understanding of the effects of mutations was shown to be relatively low, this is due typically to widespread uncertainty, rather than to misinformation; many respondents simply do not know much about (or do not know the meaning of the term) "mutation" (II-4D).

(3) As to whether certain genetic diseases are transmitted through both parents, it is uncertainty, rather than misinformation, that accounts for the lower knowledge level (II-4B, II-4C).

(4) As to the prevalence of genetic disease exclusively among males or females, both misinformation and uncertainty are moderately prevalent. Uncertainty is more widely found with reference to such disease occurrence in females than is found correspondingly for males (II-10, II-11).

Overall, in view of substantial educational attainment of the study population, rather high levels of understanding of genetic principles and of human genetic disease are anticipated and, indeed, demonstrated by the data. Even the lowest percentage of "correct answers" in the question set still

TABLE 5 Misinformation and Uncertainty Principles of Genetics and Inheritance

			Males			Females			
			20-29	*30-39*	*40-49*	*20-29*	*30-39*	*40-49*	*Total*
II-2	"inherited/ congenital"	M.	x	xx	xx	x	x	x	x
		U.	??	?	?	?	?		?
II-3	"gene definition"	M.	xx	xx	xx	xx	xx	xx	xx
		U.							
II-4A	"skip generations"	M.							
		U.							
II-4B	"inheritance/ one parent"	M.							
		U.							
II-4C	"inheritance/ both parents"	M.		x	x		x		
		U.	?	?	??	?	??	??	??
II-4D	"mutations"	M.							
		U.	??	??	??	??	??	??	??
II-9	"carrier healthy"	M.							
		U.							
II-10	"in males only"	M.	x	x	x	x	x	x	x
		U.	?	?	?	?	?	??	?
II-11	"in females only"	M.	x	x	x	x	x	x	x
		U.	??	?	??	??	??	??	??
II-12	"in subpopulations"	M.							
		U.			?			?	

"Misinformation" (M): 10.0 – 19.9% incorrect responses: x; 20% incorrect responses up: xx

"Uncertainty" (U): 10.0 – 19.9% "don't know" + no ansewer: ?; 20.0% don't know + no answer up: ??

No notation (blank spaces) in table indicate figure of *less than* 10% in designated response category.

exceeds the 50% mark—namely, at least a majority of the respondents provide accurate replies—while for five of the eleven questions, more than 90% of the respondents answer correctly, suggesting existence of a high degree of understanding of specified aspects of genetics. Within the restricted age range, age and sex differences are minor and somewhat erratic, though there is some tendency for the older respondents to be less informed than the younger. (This trend is even more evident when respondents over 50 are included; these data are not reported here.)

While general education is, no doubt, an important force in providing knowledge of genetics, this impact is not uniform for all issues. A rather "academic" item, as providing the correct definition of a "gene," is more likely to be directly linked to formal schooling, and thus is more specifically age related, with lower knowledge levels appearing at higher ages. On the other hand, that certain "public knowledge" concepts as promulgated by the mass media may transcend specific effects of academic training—for example, that certain genetic diseases occur predominantly in certain subpopulations (sickle cell anemia; Tay-Sachs)—is widely understood, as is the view that simply being a carrier of genetic disease is not in itself debilitating to a person's health.

While some experts have questioned whether the public can readily grasp the concept of probability or chance of risk in transmission of genetic disease, the data indicate that this concept provides no material difficulty to the respondents studied here. At significantly lower education levels, however, different results may occur.

While there is little doubt in the minds of the respondents that certain genetic diseases may be inherited through one parent only, considerable uncertainty seems to exist as to the possible joint role of *both* parents. This somewhat more subtle and complex precondition to genetic illness, as applies to Tay-Sachs disease, for example, requires further explanation and interpretation if it is to be understood by a greater number of persons in the potentially affected subpopulation.

PERCEIVED TREATABILITY OF GENETIC DISEASE

According to responses to the question: "Do you believe that genetic disease can be treated?" the most widely held position is that genetic disease is *sometimes* treatable (81%), while a minority (12%) believe that such disease is *always* treatable. Those who consider genetic disease as entirely *un*treatable constitute only a scattering, less than 1% (see Table 6).

As GKI levels are compared, a pattern emerges that proves characteristic of views of genetic disease manageability—for

TABLE 6 Genetics Knowledge and Genetic Disease "Treatability"

	GKI					
treatable?	I	II	III	IV	V	
always	(6) 10.2	(15) 14.0	(37) 13.0	(42) 12.3	(4) 7.4	(104) 12.3
sometimes	(37) 62.7	(78) 72.9	(231) 81.1	(291) 85.1	(49) 90.7	(686) 81.0
never	(1) 1.7	(2) 1.8	(1) 0.3	(2) 0.6	(1) 1.9	(7) 0.8
DK/NA	(15) 25.4	(12) 11.2	(16) 5.6	(7) 2.0	(0) 0	(50) 5.9
Total n %	(59) 100.0	(107) 100.0	(285) 100.0	(342) 100.0	(54) 100.0	(847) 100.0

X^2, sig.***

the criterion of treatability, as well as for other criteria considered subsequently,

> the proportion of respondents who *do not know* or provide *no answer* with reference to a specific manageability issue varies inversely with GKI level.

Perhaps not surprisingly, among persons who have little genetics knowledge, the relative proportions venturing no opinion regarding disease manageability consistently are higher than corresponding proportions for those who appear to know more about genetic principles and inheritability. However, as a subsequent summary of the "don't know/no answer" responses will show, the proportion of such manifest uncertainty is relatively lowest for the "treatability" criterion; *even those who know little about genetics display considerable faith in the ability of the medical profession to* treat *genetic disease,* presumably with some hope of success, though uncertainty is more widespread with regard to curability, predictability, and preventability of genetic disease.

TABLE 7 Genetics Knowledge (GKI) and Genetic Disease "Curability"

	GKI					
Curable?	I	II	III	IV	V	
always	–	–	(2) 0.7	(2) 0.6	–	(4) 9.4
sometimes	(31) 52.5	(67) 62.3	(203) 71.2	(246) 71.9	(27) 50.0	(574) 67.8
never	(1) 1.7	(5) 4.7	(19) 6.7	(48) 14.0	(24) 44.4	(97). 11.5
DK/NA	(27) 45.8	(35) 32.7	(61) 21.4	(46) 13.4	(3) 5.6	(172) 20.3
Total n %	(59) 100.0	(107) 100.0	(285) 100.0	(342) 100.0	(54) 100.0	(847) 100.0

X^2, sig. ***

PERCEIVED CURABILITY OF GENETIC DISEASE

In responding to the question: "Do you believe that genetic disease can be cured?" more than three-fourths of the respondents indicated a belief that genetic disease would be curable. That belief typically is erroneous. "Curability sometimes" constituted the most frequently held attitude (68%). For the total sample, less than 12% believed that genetic disease is *incurable.* However, the latter position is most widely held by persons at the highest genetics knowledge level (44.4%); *indeed, for curability the relationship between GKI level and the view that genetic disease can* never *be cured is positive (monotonic), rising consistently as GKI level rises* (see Table 7).

PERCEIVED PREDICTABILITY OF GENETIC DISEASE

In reply to the question: "Do you believe that genetic disease can be predicted?" wide agreement emerges: Nearly

90% concur that genetic disease is sometimes predictable (89.7%). Especially at GKI levels IV and V, well above 90% characterize genetic disease as a potentially predictable phenomenon (see Table 8).

PERCEIVED PREVENTABILITY OF GENETIC DISEASE

As concerns preventability: "Do you believe that genetic disease can be prevented?" some seven of ten of the respondents concur that genetic disease is preventable (74.6%). Again, a consistent direct association appears in the proportions of respondents in the several GKI levels who concur that genetic disease is preventable. The lowest proportion of confidence in such preventability appears at the lowest GKI levels, with a consistent increase in proportion as GKI level increases. At GKI level V, more than 96% take the position that genetic disease is preventable, while at GKI level I the corresponding percentage is near 40% (see Table 9).

RESPONDENT UNCERTAINTY REGARDING GENETIC DISEASE MANAGEABILITY

Table 10 shows rank orders of the respondents' *uncertainty* perceptions with reference to each of the four genetic disease manageability criteria. Such uncertainty ("don't know," "no answer") is distinct, of course, from responses revealing actual *mis*information.

As noted, the proportion of respondents displaying uncertainty with regard to genetic disease manageability declines as GKI level rises. Respondents who are well informed on matters of genetics also are likely to hold views (most often *correct* views) concerning genetic disease manageability.

For the sample as a whole, the *uncertainty perceptions* ("don't know" *plus* "no answer") *are relatively highest—about 20%—for the criteria of curability and preventability.*

TABLE 8 Genetics Knowledge and Genetic Disease "Predictability"

	GKI					
Predictable?	I	II	III	IV	V	
always	(3) 5.1	(8) 7.5	(10) 3.5	(11) 3.2	(0) 0	(32) 3.7
sometimes	(35) 59.3	(84) 78.5	(262) 91.9	(325) 95.0	(54) 100.0	(760) 89.7
never	(0) 0	(1) 0.9	(2) 0.7	(3) 0.9	(0) 0	(6) 0.7
DK/NA	(21) 35.6	(14) 13.1	(11) 3.9	(3) 0.9	(0) 0	(49) 5.8
Total n %	(59) 100.0	(107) 100.0	(285) 100.0	(342) 100.0	(54) 100.0	(847) 100.0

X^2, sig.***

TABLE 9 Genetics Knowledge and Genetic Disease "Preventability"

	GKI					
Preventable?	I	II	III	IV	V	
always	(2) 3.4	(5) 4.7	(13) 4.6	(7) 2.0	(1) 1.9	(28) 3.3
sometimes	(22) 37.3	(69) 64.5	(191) 67.0	(271) 79.2	(51) 94.4	(604) 71.3
never	(1) 1.7	(5) 4.7	(17) 6.0	(29) 8.5	(2) 3.7	(54) 6.1
DK/NA	(34) 57.6	(28) 26.2	(64) 22.5	(35) 10.2	(0) 0	(161) 19.0
Total n %	(59) 100.0	(107) 100.0	(285) 100.0	(342) 100.0	(54) 100.0	(847) 100.0

X^2, sig.***

However, at the lowest GKI level, the uncertainty proportion is highest for preventability, even beyond uncertainty respecting curability.

Regarding predictability and treatability, for the entire sample the uncertainty proportions are slightly below 6%. In each instance, it is again the lowest GKI level that reveals highest uncertainty proportions, 35.6% for predictability and 25.4% regarding treatability.

TABLE 10 Comparative "Uncertainty" Perceptions Regarding Genetic Disease Manageability Criteria

	I	*II*	*III*	*IV*	*V*	*M*
Curability	2 (45.8)	1 (32.7)	2 (21.4)	1 (13.4)	1 (5.6)	1 (20.3)
Preventability	1 (57.6)	2 (26.2)	1 (22.5)	2 (10.2)	3[a] (0)	2 (19.0)
Treatability	4 (25.4)	4 (11.2)	3 (5.6)	3 (2.0)	3[a] (0)	3 (5.9)
Predictability	3 (35.6)	3 (13.1)	4 (3.9)	4 (0.9)	3[a] (0)	4 (5.8)
unweighted "uncertainty" mean	(41.1)	(20.8)	(13.4)	(6.6)	(1.4)	(12.8)

NOTE: Arabic numerals outside parentheses indicate uncertainty ranks, considered vertically *within* specified GKI level rubric. Numerals in parentheses indicate percentages: 'don't know' + 'no answer', for specified GKI level rubric. Uncertainty percentage in excess of 20% *underlined.*

a. tie

SUMMARY AND DISCUSSION

With the aid of a convenient yardstick of genetics knowledge, the GKI, positively linked to general education, perceptions of genetic disease manageability are explored, as observed in a married, reproductive-age, highly educated Jewish population in Los Angeles.

Genetic disease is very widely regarded as predictable (sometimes, 89.7%; always, 3.7%) and as treatable (sometimes, 81.0%; always, 12.3%). To a somewhat lesser extent, genetic disease is viewed as preventable (sometimes, 71.3%; always, 3.3%), and as curable (sometimes, 67.8%; always, 0.4%).

The attitude profile, therefore, speaks strongly for the existence of a *highly optimistic* baseline concerning genetic disease manageability. For some manageability criteria, prevailing optimism may exceed the bounds of present medical reality; this is the case especially for curability. Here more than two-thirds in the sample express the belief that genetic disease is curable, though this fails to be the case currently for any genetic condition.

In a similar vein, the high proportion who regard genetic illness as treatable (one may assume with the expectation of some possible success) reveals an exceptionally widespread faith in the capability of the physician to intervene in alleviation of genetic disorders. It may be that the population studied is inclined to generalize to genetic disease attitudes conventionally held regarding disease control *in general,* without consideration of the special conditions affecting management of genetic disease. It is as though people might be saying, "After all, genetic disease is just another disease." Given this formulation, respondents are inclined to assume implicitly a usual sequence of events that may be paraphrased by the patient's statement "I don't feel well," followed by "I need a doctor," which in turn leads to a series of steps calling for medical diagnosis, treatment, and the reestablishment of well-being.

The alternate sequence of *prediction and prevention* appears less widely understood, particularly among persons with little knowledge of genetics. These mostly profess ignorance concerning genetic disease preventability. At the highest GKI levels, however (particularly at GKI levels IV and V), incurability, but also preventability, are well recognized, in line with present medical reality.

Given the pervasive assertion that genetic disease is predictable, further education efforts directed at linking such predictability with preventability may constitute an appropriate foundation for purposeful promulgation of genetic disease screening and other preventive procedures.

As noted, the population that is the focus of this inquiry stands considerably above the U.S. average in education level and, inferentially, stands high in genetics knowledge. Some hints of conditions that may be found in less-educated populations may be derived by consideration of respondents at lower GKI levels. Here the most notable finding is the consistently substantial proportion holding no clear view of crucial elements in genetic disease control. In the mounting of large-scale educational campaigns seeking voluntary participation in genetic screening programs or other similar efforts, this major opinion vacuum must be considered. Though confidence (perhaps *over*confidence) may exist regarding relatively conventional approaches to treatment and cure, attitudes toward alternate strategies of disease prediction and prevention, *instead of treatment and cure,* are less crystalized, particularly at lower genetics knowledge levels. It is only at the highest GKI level (V) that the incurability of genetic disease is fairly widely recognized, and that the view that such disease may be predicted and prevented is nearly universally affirmed. It is this combination of belief in *predictability and preventability, replacing belief in accustomed treatment and cure,* that may emerge as a key attitudinal building block in further development of communitywide genetic disease control efforts.

Chapter 4

COMMUNITY-BASED GENETICS EDUCATION

An educational effort seeking to impart knowledge of genetic principles to specific target populations or to the community as a whole inevitably is directed at people who do not come to the subject matter with a blank slate of ignorance; typically, these people have varied prior knowledge of the topic at hand, and varied reasons for wanting to know. For instance, some may be acquainted with others whose lives have been directly affected by a genetic disease, or education and training may predispose them variously to be interested in or apathetic to the issues presented. Thus, persons in a genetic education target population respond selectively to available 'messages' in the context of a number of strategic personal and sociological background factors, as, for instance were noted in Chapter 2.

The present chapter examines the relationship between sources of information on Tay-Sachs disease (communication flow), and knowledge of TSD itself.

As a criterion of TSD knowledge, we constructed an index denoted as the Tay-Sachs Knowledge Index (TSKI), based on

Authors' Note: This chapter is based on Massarik et al. (1977).

a set of five questions. This measure, with scores simply ranging from 0 to 5, is computed by noting the number of correct responses to these questions whose wording may be paraphrased as follows:

	Correct answer
"TSD symptoms evident at birth"	(no)
"TSD found only in Jewish children"	(no)
"TSD is rare, even in Jews"	(yes)
"TSD is always fatal"	(yes)
"TSD risk exists if *both* parents are carriers"	(yes)

As to communication flow, three major categories of information source, or *communication channels,* are considered:

(1) Mass Media
  Television
  Radio
  Magazines
  Newspapers
(2) Institutional Sources
  Physician
  Clergyman
  School
  Scientific literature
(3) Social Relations
  Has known family in which Tay-Sachs disease has occurred
  Relatives
  Friends

In addition, it will be useful to construct a heuristic index of *total communication flow,* to indicate the extent to which the respondent reports exposure to a relatively rich or meager information environment. These measures are then conceptualized in a genetic-information transmission schema, which may prove useful not only in development of TSD community-based education programs, but also in general dissemination of knowledge on genetic disease and its control.

TAY-SACHS DISEASE KNOWLEDGE LEVELS

As baseline in the assessment of impact of community-based education on TSD and in the study of the role of the various communication channels, we consider the overall prevailing TSD knowledge level, as indicated by the TSKI. TSKI measurements, as reported here, were taken prior to implementation of a major communitywide education program on Tay-Sachs Disease and its control, instituted subsequently by the California Tay-Sachs Disease Prevention Program. However, a television program, *Marcus Welby, M.D.*, perhaps a year earlier, had provided a large public with some knowledge of the disease.

As was the case for the GKI, we cannot judge at this time whether TSD knowledge in the respondent population is to be considered normatively "high" or "low." However, it is likely that, given the study sample's high education level, scores appearing here may compare favorably with those likely characterizing less educated population groups. Noting incidence of greater risk in Jewish populations, one may expect still lower TSKI scores in non-Jewish groups.

We find that, on the average, male respondents reply correctly to *just less than two* of the five questions ($\overline{M} = 1.75$), and females to *somewhat more than two* ($\overline{M} = 2.33$). On an *absolute* basis, it would appear that the prevailing knowledge level concerning Tay-Sachs disease is not particularly high.

Of the males, 31.1% and 11.2% of the females report that they have never heard of Tay-Sachs disease, a contrast both statistically significant and expected in view of TSD's special relevance to the Jewish woman.

A significant association between general education level and generic genetics knowledge has been previously observed. While it may be hypothesized that education will also play a role in determining the level of knowledge concerning Tay-Sachs disease, this relationship may be expected to be less distinct: After all, Tay-Sachs disease represents a considerably more specific phenomenon than genetics information generally, as normally imparted in various high school and university courses and as may appear in semitechnical and popular literature. Further, there is the issue of personal relevance; while certain facts about genetics may be of wide interest, the specific import of knowledge of Tay-Sachs disease may be greater for women—particularly for women of childbearing age who are Jewish and who may be concerned with the specific personal risk of giving birth to a Tay-Sachs child. The male partner is, of course, also affected by the potential birth of the diseased child, but he is, at any rate, one step removed from the *most* immediate and painful involvement with a Tay-Sachs birth.

In this context, we may examine for its intrinsic meaning and as basis for further analysis the general education levels and TSKI scores that characterize the (mostly) Jewish men and women, mainly of reproductive age, contacted in the Los Angeles study sample.

Tables 11 and 12 show the relationship between general education and Tay-Sachs disease knowledge for males and females, respectively. A summary of comparisons, contrasting the sexes with reference to education and TSKI appears in Table 13.

It is clear that, as expected on the basis of numerous Jewish population study results, the proportion of men with advanced education considerably exceeds the corresponding proportion for women. Just less than half (45.3%) of the men have attained advanced academic degrees, compared to about one woman in seven (13.8%).

A net comparison of general education, based on a heuristic index (see Table 13), shows the higher overall education

TABLE 11 General Education Level and Tay-Sachs Disease Knowledge (% by Education)

	Males				
	high school or less	*some college*	*college graduate*	*professional degree*	*Total*
TSKI					
0	47.4	37.5	28.9	27.0	30.2
1	26.3	25.0	15.6	19.0	19.0
2	10.5	23.2	18.5	19.0	19.0
3	15.8	8.9	23.0	12.1	15.6
4	0	3.6	7.4	13.2	9.1
5	0	1.8	6.7	9.8	7.0
Total %	100.0	100.0	100.0	100.0	100.0
Number of cases	19	56	135	174	384
Education distribution (%)	4.9	14.6	35.2	45.3	100.0

level of the men (college grad +, 3.21), as contrasted with that of the women, (some college +, 2.59).

With the above background regarding general education, we note important differences between the sexes in knowledge levels for Tay-Sachs disease, considerably beyond differences attributable to general education impact. A review of information shown in Tables 11 and 12, and highlighted in summary form in Table 13, makes it clear that *at all education levels, females know more about Tay-Sachs disease than do men with corresponding education.* For instance, among those with high school education or less (rather small subgroups of the sample), nearly one-half (47.4%) of the males

TABLE 12 General Education Level and Tay-Sachs Disease Knowledge (% by Education)

	Females				
	high school or less	*some college*	*college graduate*	*professional degree*	*Total*
TSKI					
0	20.0	13.7	11.5	15.5	13.6
1	14.3	19.9	15.2	12.1	16.5
2	22.9	19.9	20.0	29.3	21.5
3	28.6	24.8	26.7	25.9	26.0
4	8.6	18.6	18.9	13.8	17.2
5	5.7	3.1	7.9	3.4	5.3
Total %	100.0	100.0	100.0	100.0	100.0
Number of cases	35	161	165	58	419
Education distribution (%)	8.4	38.4	39.4	13.8	100.0

did not provide a single correct answer on the five TSKI questions; while the corresponding figure for women is only 20%. Even among those with college degrees or advanced professional training, the percentage of men providing no correct answers is sizable (ranging from 27 to nearly 29%).

The male/female gap in TSKI scores is most extreme at the "high school or less" education level, but it becomes less distinct with increasing education; education seems to make men and women more equal in TSD knowledge. Still, women appear to be favored throughout: For those with high school education or less, the women score 2.09 (TSKI correct answers), compared with the males' score of less than 1.00 (0.95 TSKI correct answers). At the professional degree level,

TABLE 13 Comparisons; Males/Females: General Education Level TSKI Level

	Males	*Females*
General Education Level (index):[a]	3.21	2.59
TSKI Level (total):[b]	1.75	2.33
high school or less	0.95	2.09
some college	1.22	2.24
college grad.	1.85	2.50
professional degree	1.95	2.21

a. high school or less = 1, some college = 2, college graduate = 3, professional degree = 4

b. average number of correct ansewers to TSKI questions, e.g. 0 = no right ansewer (max. possible = 5)

the corresponding scores are 1.95 for males and 2.21 for females, a difference of evidently lesser magnitude.

These findings lend credence to the assertion that Jewish women are considerably more knowledgeable with regards to Tay-Sachs disease than are Jewish men. Men, on the other hand, knowing less about this disease than women at all education levels, appear to become *relatively* more knowledgeable as their formal education increases.

The lower dependence of women's TSKI scores on education suggests the prevalence of a higher intrinsic concern on their part, with the potential direct threat of Tay-Sachs disease in childbearing, and with other familial genetic risk. However, this finding must be seen against the background of the means of communication associated with general "publicity" and community-based education efforts, and with differentially active search for Tay-Sachs disease information by the men and women included in this study sample. Next, therefore, we shall consider communication flow—the several information channels from which TSD knowledge is derived, and their association with various levels of TSD knowledge among male and female respondents.

TAY-SACHS DISEASE KNOWLEDGE INFORMATION FLOW

As shown in Table 14, television emerges as the largely unchallenged champion, as the "medium of choice" from which TSD information is obtained. Television ranks number one for both males and females, virtually across the entire spectrum of TSKI levels with exception of the lowest. The impact of TV may be traced in part to a widely-heralded TV drama on the topic.

Generally, newspapers—though they rank respectively third for females and second for males—tend to decline on a relative basis as a TSD information source as TSKI knowledge level *rises.* (However, this decline is relative; absolute percentages are high across the spectrum of TSKI scores.)

Magazines stand in third and fourth place, respectively, among media for males and females. A significant contrast appears in relation to friends as a TSD information source: *Women, in higher proportion than men, report that they have obtained TSD information through this social relationship, rather than by mass media.*

It is noteworthy that the role of the rabbinate is virtually nil as a practical source of TSD information. Further, with exception of the highest knowledge level, particularly for men, *the impact of the physician as an information source is modest indeed.* These lacks suggest that these communication channels are not satisfactorily utilized at present, and that they may be well mobilized more effectively in future community-based genetics education efforts.

At the very highest level of TSD knowledge, for males certain specialized sources, such as scientific literature rank highest as information source. It would appear that those males who obtain perfect scores in the TSKI represent a rather specialized population with considerable interest and technical knowledge in the field.

TABLE 14 Ranks: Most Frequently Reported TSD Communication Channels (first four ranks shown only)

	Total		*TSKI* *0*		*1*		*2*		*3*		*4*		*5*	
	M	F	M	F	M	F	M	F	M	F	M	F	M	F
Television	1	1	2	3	1	1	1	1	1	1	1	1	2	1
Magazines	3	4	3	2	3	3	3	3	2.5	4	3	3	-	3
Newspapers	2	3	1	1	2	2	2	4	4	3	4	4	4	4
Physician	-	-	-	-	-	-	-	-	-	-	-	-	3	-
School	-	-	4	-	-	-	-	-	-	-	-	-	-	-
Sci. Lit.	-	-	-	-	-	-	-	-	-	-	2	-	1	-
Friends	4	2	-	4	4	4	4	2	2.5	2	-	2	-	2

Tables 15 and 16 show the association between TSD communication channels and TSD knowledge level (TSKI) for men and women. These data, form which the ranks shown in Table 14 were generated, affirm the absolute dominant position of television (perhaps by the *Marcus Welby, M.D.* program on TSD particularly) as a major source of information.

The typically high level of reliance on friends among females as an information source, is of particular interest. No matter what the TSKI level, the proportion of women reporting that they have obtained TSD information from friends exceeds the corresponding proportion for men. On the other hand, the latter more typically report having obtained their information from the scientific literature and from school, though both sexes tended to respond to mass media in almost equal measure. Aspects of these results are highlighted in Table 17.

Table 17, taking account *only* of those who have heard of TSD, according to their questionnaire response, reveals a virtually identical Mass Media Index (31, 32), while the Institutional Index, including information sources through school, scientific literature, and physicians, is higher for men

TABLE 15 Association between TSD Communication Channels and TSD Knowledge Level (% of respondents reporting that TSD information obtained from specified communication channel)

	Males						
	TSKI						
	0	*1*	*2*	*3*	*4*	*5*	*Total*
Television	7.8	31.5	47.9	47.6	47.1	40.7	34.6
Radio	0.9	12.3	11.0	1.7	5.7	7.4	6.0
Magazines	6.9	24.7	30.1	23.7	37.1	29.6	21.7
Newspapers	8.6	28.8	34.2	22.0	34.3	33.3	23.6
Physician	0	2.7	6.8	11.9	28.6	37.0	8.9
Clergyman	0	0	2.7	0	0	0	0.5
School	2.6	2.7	1.4	8.5	25.7	29.6	7.3
Sci. Lit.	0	4.1	5.5	15.2	48.6	51.9	12.3
TSD Family	2.3	1.4	8.2	15.3	31.4	29.6	9.9
Relatives	0.9	8.3	12.3	11.9	8.6	18.5	8.1
Friends	1.7	13.7	27.4	23.7	22.9	29.6	16.2
have *not* heard of TSD	80.0	24.7	6.8	6.8	0	0	31.1
total no. of cases	115	73	73	59	35	27	382[a]
% of cases	30.1	19.1	19.1	15.4	9.2	7.1	100.0

NOTE: Percentages read vertically and do not necessarily add to 100, because of multiple media mentions. Totals, not shown, would appear just above total no. of cases.

a. 2 cases with missing data excluded

than for women (11, 5). On the other hand, as observed, the Social Relations Index, noting information sources such as acquaintance with a TSD family, relatives, and friends, tends to be relied on in higher proportion by women than by men.

As the *total* information flow is calculated for those who *have* heard of TSD (the CC Index), the differences between men and women appear to be nil: Rather, it is a matter as to *which* information sources are relied on more heavily as basis for acquisition of TSD knowledge. (But we must recall that fewer men than women have heard anything at all about Tay-Sachs disease; those who have not heard of TSD do not appear in the data reported in Table 17.)

TABLE 16 Association between TSD Communication Channels and TSD Knowledge Level (% of respondents reporting that TSD information obtained from specified communication channel)

	Females						
	TSKI						
	0	*1*	*2*	*3*	*4*	*5*	*Total*
Television	12.3	47.8	51.1	52.3	65.3	72.7	49.2
Radio	1.8	4.3	3.3	2.8	2.8	4.5	3.1
Magazines	14.0	26.1	33.3	36.7	31.9	50.0	31.0
Newspapers	19.3	31.9	31.1	39.4	27.8	45.5	32.0
Physician	0	5.8	4.4	5.5	11.1	40.9	7.4
Clergyman	0	0	2.2	0.9	0	0	0.7
School	1.8	0	5.6	4.6	1.4	0	2.9
Sci. Lit.	1.8	5.8	2.2	9.2	8.3	31.8	7.2
TSD Family	0	7.2	12.2	22.0	20.8	31.8	14.8
Relatives	1.8	8.7	8.9	9.2	13.9	18.2	9.3
Friends	7.0	21.7	42.2	43.1	37.5	54.5	34.1
have *not* heard of TSD	57.9	10.1	2.2	3.7	1.4	0	11.2
total no. of cases	57	69	90	109	72	22	419
% of cases	13.6	16.5	21.5	26.0	17.2	5.3	100.0

NOTE: Percentages read vertically, and do not necessarily add to 100, because of multiple media mentions. Totals, not shown, would appear just above total no. of cases.

A CONCEPTUAL FRAMEWORK FOR THE STUDY OF TSD/GENETIC-INFORMATION TRANSMISSION

The various media of communication imparting knowledge of general genetics and of specific genetic disease must be viewed as embedded in a context of personal and sociological variables, noted in the schema, Chapter 2. This "field" of influences represents a complex interplay of forces that, in their eventual operation, lead to greater or lesser levels of TSD knowledge and, indirectly, to various levels of compliance in genetic screening. A conceptual framework, relating to transmission of genetics information, as illustrated by TSD knowledge, appears in Figure 3.

A basic contextual variable, well-explicated in behavioral science literature, may be designated *life-status.* It denotes a pattern of variables of which, in the present instance, the

TABLE 17 For Males and Females: TSD Communication Channels Reported *Only* by Those Who Have Heard of TSD

	Total	
	Males	*Females*
Television	50.2	55.4
Radio	8.7	3.5
Magazines	31.5	34.9
Newspapers	34.2	36.0
Mass Media: Index	31	32
Physician	12.9	8.3
Clergyman	0.7	0.8
School	10.6	3.3
Sci. Lit.	17.9	8.1
Institutional: Index	11	5
TSD family	14.4	16.7
Relatives	11.8	10.5
Friends	23.5	38.4
Social Relations: Index	17	22
CC Index	20	20
% have heard of TSD	68.9	88.8

NOTE: Indexes constructed by summing individual percentages within specified media groups and dividing sum by number of media constituting the media group.

following are illustrative: age, sex, marital status, and Jewish identification. The present research holds constant, within specified limits, several of these (themselves interrelated) variables. Thus, in most analyses, the age range is restricted to the reproductive ages, and the sample significantly represents fairly explicitly identified Jews in Southern California. While some differences related to age appear, even within the constrained range, relevant contrast may be predicted (or postdicted) on the basis of the sex variable, considered within the frame of life-status.

Cultural differences persist, of course, between the sexes in their principal role assignments, in spite of the feminist movement and alternative lifestyles. With the male heavily oriented toward professional achievement, typically via formal academic education (notably so in the Jewish population), and with the female oriented significantly toward home and childrearing, different attitude sets toward genetics issues may be anticipated. Thus, flowing from the life-status variable, we distinguish, for men, the strategic influence of *general education* and, for women, a *special concern* with genetics/TSD as it relates to childbearing. The latter variable, psychologically more closely linked to issues of TSD knowledge and genetic screening, may be regarded as the more potent "active ingredient" in establishing predisposition toward higher motivation to learn/know about Tay-Sachs disease, than general education. However, education itself incorporates a certain amount of general knowledge of genetics (particularly as education level rises), though not necessarily in equal measure knowledge of Tay-Sachs disease—a far more specific topic of personal impact in a given subpopulation.

An additional salient variable relates to *direct experience* with TSD. Such experience may be deemed an important incentive for seeking further information concerning the disease, particularly if other conditions, rooted in the life-status variable (e.g., Jewish identification, perception of TSD as a "Jewish disease," and reproductive age) are met.

The potent impact of special concern, as is more inclined to be manifest by the female (though the male may share this concern in lesser measure), is likely to establish a *high* level of predisposition toward learning about this disease. The male, on the other hand, whose relatively most salient variable may be general education, as associated with professional activity, rather than the woman's special concern, may be characterized only by a corresponding *moderate* predisposition.

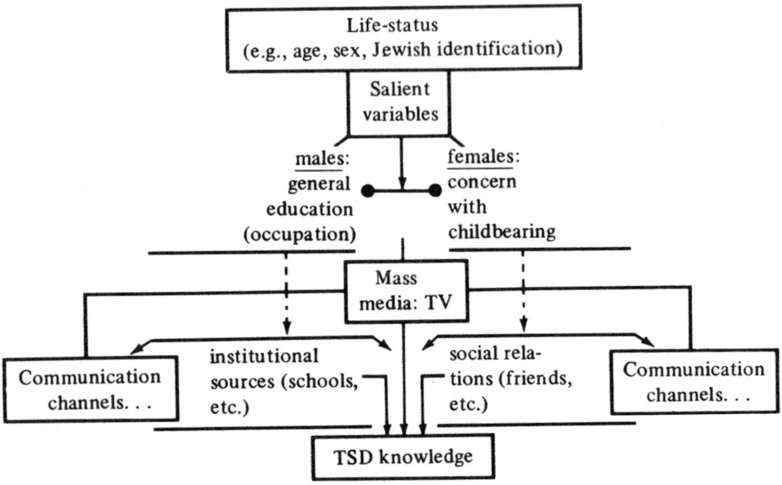

Figure 3: A conceptual framework for the study of the transmission of information on genetics. Life-status determines sets of Salient Variables that differ, on cultural grounds, between males and females. These specific Salient Variables in turn determine differentially used and differentially effective Communication channels. However, Mass media tend to influence both males and females. Thus, TSD knowledge is a joint product primarily of specific Life-status-linked variables, leading to selectivity in Communication channels, and of Mass media impact.

The mass media provide general information input and may play a supportive and stimulating role, but do not necessarily generate a crucial differentiating effect between men and women in establishing TSD knowledge levels. The impact of the special concerns, such as those motivating further search for information by the female, and particular educational-professional interests of the men may make the difference between higher and lower TSD knowledge levels among males and females.

In view of the higher psychological potency of the female's special concerns, we would hypothesize that search for additional information and "reality testing" of views concerning TSD would be greater for women than for men. This may be indicated by the function of information transmission through friends, a process much less characteristic of the men studied. On the other hand, deriving additional knowledge from the scientific literature and from schools is more typical of men as compared with the women in the study sample. Again, we must consider the possible impact of changing lifestyles and social values on these differential modes of utilizations of knowledge sources.

We may conclude that, while mass media do provide significant amounts of information concerning Tay-Sachs disease—and while television is by far the most pervasive medium and information source—the operation of *specific concerns directly linked to life-status and aspects of continuing sex-specific roles* is important in establishing Tay-Sachs disease knowledge levels and, perhaps, in consolidating such knowledge as is eventually obtained.

Further, one must recall a possible connection between *specific choice* of mass media and education and TSD knowledge level, with particular media favored at lower levels, and with mass media as a whole playing a relatively greater role at the lower knowledge levels than institutional sources and social relations, whose impact rises with increasing knowledge.

Given these considerations, in accordance with the data-based framework proposed, we would be inclined to predict that, in future research, TSD knowledge levels of females, ceteris paribus, will be shown to exceed such levels for men of corresponding age, unless specific community-based education programs take explicit account of the *men's relatively lower predisposition to learn about Tay-Sachs disease.*

It does not necessarily follow from this framework that TSD knowledge and compliance with specific genetic screening programs are clearly synonymous. Though information provides an important starting point for action, the role of peer and reference groups (such as the friends included among the communication channels), geographical location, and aspects of mood and near-term swings in motivation must be considered, as suggested in the Chapter 2 Schema, if we are to establish a more rigorous model of compliance in genetic screening. In this framework, as one approaches planning of community-based genetics education, it is important not to assume the existence of a homogeneous, complacently receptive public, just waiting to be "educated." Rather, one needs to assess the interplay of preexisting circumstances, notably the differential life-status positions and the special concerns and individual motivations of the persons to be educated, in the context of their salient social and personal worlds.

Table 18, the Tay-Sachs Disease Information Matrix, addresses the question: "To what extent are the claims by respondents that they have heard more or less of Tay-Sachs disease borne out by their actual knowledge of the disease, as measured by the Tay-Sachs (Disease) Knowledge Index? Further, the matrix provides an indication of the extent to which, in accordance with the respondents' testimony, they have or have not been significantly exposed to information regarding this ailment.

With regard to the latter inquiry, it appears that considerably more men than women (31.2 versus 11.2%) note that they have *never* heard of Tay-Sachs disease. At the other extreme, somewhat more women than men (16.0 versus 10.5%) assert that they have heard *a lot* about the disease. The remainder (58.4% of the males and 72.8% of the females) note that they have heard "something" about this illness. These findings confirm the assumption that women, in larger proportion than men, view themselves as knowledgeable with reference to the disease, modest proportions at the highest level ("a lot"), but an overwhelming majority at the level of having heard "something" concerning the matter.

As one notes, the distribution of TSKI scores for the various levels of what we may call "perceived knowledge" (viz. "having heard of TSD," as against "tested knowledge" equivalent to the TSKI), show considerable differences. Among those who report that they have never heard of TSD

TABLE 18 The Tay-Sachs Disease Information Matrix perceived knowledge = "heard of TSD?"

tested knowledge = TSKI	*never*		*something*		*a lot*	
	M	*F*	*M*	*F*	*M*	*F*
0	77.3 (92)	70.2 (33)	10.3 (23)	7.2 (22)	0 (0)	3.0 (2)
1	15.1 (18)	14.9 (7)	24.7 (55)	19.7 (60)	0 (0)	3.0 (2)
2	4.2 (5)	4.3 (2)	30.0 (67)	26.6 (81)	2.5 (1)	10.5 (7)
3	3.4 (4)	8.5 (4)	21.1 (47)	26.6 (81)	20.0 (8)	35.8 (24)
4	0 (0)	2.1 (1)	9.9 (22)	17.7 (54)	32.5 (13)	25.4 (17)
5	0 (0)	0 (0)	4.0 (9)	2.3 (7)	45.0 (18)	22.3 (15)
	100.0 (119)	100.0 (47)	100.0 (223)	100.0 (305)	100.0 (40)	100.0 (67)
% who heard of TSD:	31.2	11.2	58.4	72.8	10.5	16.0
TSKI score (M)	0.34	0.57	2.08	2.35	4.20	3.45

NOTE: consistency at extremes;

(never) – (TSKI = 0.1)	males	110/382 = 28.8; 110/119 = 92.4
	females	40/419 = 9.5; 40/ 47 = 85.1
(a lot) – (TSKI = 4.5)	males	31/382 = 8.1; 31/ 40 = 77.5
	females	32/419 = 7.7; 32/ 67 = 47.8

only very low mean TSKI scores appear, 0.34 for men and 0.57 for women; thus, even those women who claim to have never heard of the disease in fact may have somewhat greater knowledge of it than men making comparable statements. When the reported level of perceived knowledge is in the middle, as "something," again the women's score exceeds that shown for the men: the latter with a mean of 2.08, and the former with an equivalent score of 2.35. However—and perhaps surprisingly—at the level at which the highest claim of perceived knowledge appears—e.g., those noting that they have heard "a lot" about TSD, the pattern reverses: Now men demonstrate a higher knowledge level, 4.20, versus the women's score of 3.45. This result may suggest the possibility that a disproportionate number of women sense, explicitly or implicitly, that they "ought to" know something of a disease, whether or not they do so in actuality, thus increasing the likelihood that they will overstate their acquaintance (and presumed knowledge level) as regards TSD. Thus, there are somewhat fewer women who provide consistent responses at the high end of the scales (see the four cells in the lower right-hand quadrant of the matrix), as contrasted with those in the upper left-hand corner. Putting it differently, those who have little actual "tested knowledge" (e.g., low TSKI scores) more typically admit that they have never heard of the disease; on the other hand, there are higher proportions than may be expected, particularly among the women, whose "tested knowledge" is toward the middle range, rather than at the top, who nonetheless claim that they have heard a lot about the disease.

It may be useful to examine, in future research, the characteristics of those persons who are inclined to overstate the degree of their knowledge of this or of other diseases that—for medical and personal reasons—may be especially significant and, perhaps, potentially threatening.

Chapter 5

DOES IT REALLY MATTER?: A Note on Prevalence and Perceived Prevalence of Genetic Disease

Human life may be priceless. Still, if a particular order of diseases appears with very low frequency, and if people at the grass roots and at decision-making levels regard the numbers as insignificant (at least as compared with other more pervasive and more detrimental conditions), then certain ailments may receive little by way of resources for their amelioration and control. The issue must be raised, therefore, at two levels:

- "What is the 'actual' prevalence of genetic disease?"
- "What is the *perceived* prevalence of genetic disease?"

Even casual reflection will indicate that questions such as these cannot yield precise answers. It is often noted that genetic disease *as such* is rare. However, estimated frequencies vary for various kinds of genetic diseases. And the *perception* of such disease frequency may be thought of as a *joint product* (a) of "actual" frequencies, per scientific estimation *and* (b) of the process of education and interpretation (as by the media) focusing on the specified genetic condition. Though (a) may be numerically significant and of considerable social impact, (b) may, in some instances, ignore the real-world condition, resulting in a dearth of allocable resources or medical compliance with procedures that may serve to control the condition. Conversely, great rarity (and perhaps slight impact) of the condition at (a), but high public dramatization of the illness, may be viewed as constituting notable societal malfunction, resulting in disproportionate resource allocation, as compared with other priorities. Finally, a suitable balance may be sought in which the prevalence, and individual and social significance, per (a), is reflected in (b) by appropriate public attention and resulting resource allocation. It is evident that the precise parameters of (a) are subject to the difficulties of formal identification and measurement, while those of (b) are affected by the human virtues and vices of partisanship and argumentation. However, this issue of relative equilibrium between "what's out there" and "what people think is out there" initially may be examined conceptually, as a basis for analysis and policy formation.

The following estimates of appearance of various kinds of genetic disorders are of interest:

> In the United States, 100,000 to 150,000 infants are born each year—*three to five percent* of . . . annual births with a significant congenital malformation . . . chromosomal abnormality . . . or a clearly defined "genetic disorder". Hereditary diseases and congenital defects account for at least one-fifth of all infant deaths in our country [National Institutes of Health, 1979: 236].

Elsewhere Scriver et al. (1978: 947; italics in original) note that

> frequency [of genetic disease] varies with geographic region and ethnic group. . . . Chromosomal mutations occur in about five of every ten spontaneously aborted fetuses, in about five percent of

> stillbirths and of infants dying within seven days of birth, in about five per one thousand surviving live births; *about ten percent of live-born infants have some form of genetic variation that could be responsible for handicap during the person's lifetime.* . . . At least twelve percent of adult hospital admissions reflect disease with a significant genetic component.[1]

Thus no single figure can answer the first question of "actual frequency." Whatever, some ballpark summary figures might hover between 5 and 10% (excluding congenital malformations) when various definitions and analyses of the medical evidence as a whole are considered.

If we accept the 5 to 10% range as being generally "correct" with reference to an acceptable (if not entirely rigorous) proportion of genetic causation of illness, we find that just less than one-half of the study population hold a generally "correct" view of genetic disease incidence (see Table 19). At the respective extremes, some 12% view genetic illness as so rare as to constitute less than 1% of all illness causation, while about 13% believe that one-half or more of all diseases are predominantly due to genetic factors.

Interpreting the pattern of response, one must keep in mind the high degree of ambiguity inherent in the phenomenon. While in strictly technical terms one may define illness predominantly as due to inherited factors or genetic causes within certain ranges, the interaction with situational and environmental factors heightens the uncertainty for certain disease types (e.g., some kinds of cancer? heart disease?).

The study data point to two conclusions:

- A sizable majority of respondents (more than 60%) indeed regard genetic cause of illness as quite rare–namely, as occurring in 10% or less of all illnesses. On the other hand, it is noteworthy that some 12 to 13% appear convinced that *50%* or more of all illnesses stem from genetic factors.
- The respondents *do* have opinions–right or wrong (indeed, in large measure right)–regarding this complex and ambiguous mat-

ter. (Elsewhere high don't know/no answer figures show respondents' readiness to admit their lack of opinion, when applicable.)

These conclusions suggest that the respondents approach the issue of genetic disease and its control with an articulated belief background, one that does not suggest flight into refusal to address the issue, nor one that is dramatically off target. Though undoubtedly some guesswork and uncertainty prevail as people view this quite complicated issue of genetic disease prevalence, they demonstrate as well reasonable perspective and probable readiness for further learning regarding genetics and associated disorders.

TABLE 19 Perceived Incidence of Genetic Disease

(Q II-1)

Percentage of all illness almost entirely/predominantly due to inherited factors/ genetic causes	*% of Respondents*	*Number of Cases*
< 1%	11.7	(99)
5%	24.7	(209)
10%	24.6	(208)
25%	23.1	(196)
50%	10.6	(90)
>70%	2.1	(18)
N.A. D.K.	3.2	(27)
	100.0	(847)

"While illness can have many causes (infection, trauma, malignancy), certain diseases are due *almost entirely* to inherited factors. What percentage of all illnesses do you think are predominantly due to inherited factors (genetic causes)? less than 1%; 5%; 10%; 25%; 50%; 70%; or more."

As one explores a possible relationship between Genetics Knowledge Index (GKI) levels and views of prevalence of genetic disease, few differences appear (see Table 20).

It is shown that genetics knowledge and perceived genetic disease prevalence are not systematically associated, though a minor (statistically marginal) tendency toward a curvilinear association appears: Among persons at the lowest GKI level (level I), 54% view the disease etiology noted at "5% or less" prevalence. The corresponding figure at mid-range, GKI level III, is lower, near 38%, while in the highest category, GKI level V, the proportion rises once more, to 48%. Thus, both people who know little and those who know a lot about genetics share in common a somewhat more widespread belief in the rarity of genetic etiology of disease. (No such indication of curvilinearity appears at the high end [50% and up] of the perceived prevalence scale.)

TABLE 20 Genetics Knowledge (GKI) and View of Genetic Disease Prevalence

	GKI					
Prevalence	*I*	*II*	*III*	*IV*	*V*	
1%	(11) 21.2	(6) 6.1	(33) 11.8	(41) 12.2	(8) 14.8	(99) 12.1
5%	(17) 32.6	(24) 24.2	(73) 26.1	(77) 23.0	(18) 33.3	(209) 25.5
10%	(12) 23.1	(29) 29.3	(75) 26.8	(82) 24.5	(10) 18.5	(208) 25.4
25%	(8) 15.4	(22) 22.2	(66) 23.6	(90) 26.9	(10) 18.5	(196) 23.9
50%	(3) 5.4	(16) 16.2	(27) 9.6	(39) 11.6	(5) 9.3	(90) 11.0
70%, up	(1) 1.9	(2) 2.0	(6) 2.1	(6) 1.8	(3) 5.6	(18) 2.2
Total n %	(52) 100.0	(99) 100.0	(280) 100.0	(335) 100.0	(54) 100.0	(820) 100.0

X^2, n.s.

NOTE: 27 nonrespondents excluded in this analysis

As a general trend, respondents are inclined to regard disease arising from genetic causes as clearly uncommon. However, there is considerable variation in perception of genetic disease prevalence: Near one-quarter in each instance hold that some 10%, or 25%, respectively, of all illnesses are genetically caused. Accordingly, information campaigns concerned with expansion of a relevant knowledge and attitude base supportive of genetic disease prevention programs also must consider this wide dispersion in perception of prevalence, as well as the dominant view asserting the relative rarity of genetic disease.

Another view of the perceived significance of genetic disease may be obtained by examining the respondents' perceptions of various illnesses . . . what diseases are regarded as being caused primarily by genetic factors? The results appear in Table 21.

Three of the diseases included—Tay-Sachs disease, sickle cell anemia, and diabetes—are widely recognized as of genetic origin, even among those whose gentics knowledge is quite low (GKI level I). In these instances, at least one-half of the respondents are aware of genetic etiology, with quite consistent increases in this awareness as genetics knowledges level rises.

For three other diseases—dwarfism, color blindness, and mongolism—there is little doubt as to a genetic basis on the part of those who are generally knowledgeable regarding genetics, particularly at GKI levels IV and V. While the pattern of cause may be more complex, there is a quite widespread perception of allergy and asthma as being principally rooted in genetics. Near sightedness also receives fairly high mention in this regard, especially at the highest GKI level.

The pattern for cystic fibrosis differs dramatically. While very high proportions of those most knowledgeable (GKI level V) recognize this disorder as genetic, only a few among all others (i.e., at GKI levels I-IV) share this recognition. It

TABLE 21 Disease Perception: Percentage Regarding Disease as "Predominantly Genetic" and Genetics Knowledge (GKI)

	For diseases medically classified as genetic					
			GKI			
	1	*2*	*3*	*4*	*5*	*trend*
Tay-Sachs Disease						
M	42	78	79	95	100	rises w/GKI, consistent
F	59	85	93	98	100	rises w/GKI‘ consistent
Sickle Cell Anemia						
M	55	85	95	99	100	rises w/GKI, consistent
F	59	83	97	98	100	rises w/GKI, consistent
Diabetes						
M	55	45	66	83	94	rises w/GKI, one 'drop'
F	68	75	82	89	94	rises w/GKI, consistent
Dwarfism						
M	32	38	55	84	100	rises w/GKI, consistent
F	37	47	55	76	89	rises w/GKI, consistent
Color Blindness						
M	19	48	66	84	100	rises w/GKI, consistent
F	18	33	58	91	100	rises w/GKI, consistent
Mongolism						
M	19	33	41	68	91	rises w/GKI, consistent
F	27	22	34	59	89	rises w/GKI, one 'drop'
Cystic Fibrosis						
M	3	15	11	25	82	rises w/GKI, one 'drop'
F	0	8	18	29	83	rises w/GKI, consistent
	Other diseases (including genetic and mixed etiologies)					
Allergy						
M	42	38	59	70	73	rises w/GKI, one 'drop'
F	23	70	72	70	72	no trend
Asthma						
M	36	30	49	57	67	rises w/GKI, one 'drop'
F	23	62	53	58	44	no trend
Near-sightedness						
M	19	33	34	49	67	rises w/GKI, consistent
F	18	35	44	56	56	rises w/GKI, consistent

TABLE 21 *(continued)*

	Other diseases (including genetic and mixed etiologies)					
			GKI			
	1	*2*	*3*	*4*	*5*	*trend*
Heart Disease						
M	36	33	44	43	33	no trend
F	14	43	45	48	39	no trend
Kidney Disease						
M	7	8	18	24	24	rises w/GKI
F	9	20	23	21	33	rises w/GKI minor trend
Leukemia						
M	7	10	19	23	24	rises w/GKI
F	0	8	9	17	28	rises w/GKI
Cancer						
M	13	5	14	15	33	no trend
F	14	13	21	20	17	no trend
Mental Illness						
M	19	23	25	29	36	rises w/GKI, consistent
F	32	25	28	21	17	*declines* w/GKI, one 'rise'
	Nongenetic diseases					
Rheumatic Fever						
M	3	10	6	10	3	no trend
F	5	2	3	4	6	no trend
Gonorrhea						
M	7	3	0	1	0	declines w/GKI?
F	9	5	5	0	0	declines w/GKI?
German Measles						
M	0	5	3	0	0	no trend
F	0	2	0	0	0	no trend
Pneumonia						
M	0	0	0	0	0	nil
F	0	0	0	0	0	nil

NOTE: Perception by 50% or more as genetic, underlined.
Percentages rounded.

appears that the public image of cystic fibrosis as a genetic ailment is not clearly communicated, even to the generally well-educated cross section of Jewish respondents included in this study.

A number of other diseases, including those accounting for high proportions of all deaths–heart disease and cancer, as well as kidney disease and leukemia–show mixed perceptions regarding corresponding impact of hereditary forces. For instance, for heart disease, although there is no trend relating to GKI level, nearly 40% ascribe genetic cause to an illness so classified. For cancer, comparable percentages are somewhat lower, generally in the 10 to 20% range, though 33% of males at the highest GKI level ascribe cancer to genetic causes.

For mental illness, an unusual and intriguing response profile appears: While for males the proportion assuming some genetic association rises–a quite frequent pattern throughout the analysis–for females the *reverse* is true; among women, those who know least about genetics are most likely to assume that mental illness is genetically caused. While an interpretation of this finding is not self-evident, one may speculate that a variety of "old wives' tales," superseding either specific knowledge or lack of opinion, is responsible for this discrepant result.

Finally, the study examined a set of clearly nongenetic diseases: rheumatic fever, gonorrhea, German measles, and pneumonia. With only minor exceptions, the study respondents virtually without regard to GKI level were cognizant of the nongenetic character of these illnesses.

Reviewing the several constellations of disease perception, with respect to imputed genetic origin, the following may be distinguished:

- Illnesses widely recognized as genetic, even by the less knowledgeable–Tay-Sachs disease, sickle cell anemia and diabetes. However, in spite of this quite wide recognition, the likelihood that these diseases will be correctly identified as genetic rises consistently as genetics knowledge (GKI level) rises, roughly with a recognition level beyond the 50% mark for the least knowledgeable, moving to almost complete recognition among the most knowledgeable.

- Those illnesses that are correctly seen by the most knowledgeable as genetically caused, while those who are less aware of genetic facts considerably lag behind the others in this respect. For these diseases, special educational efforts—even beyond any more usual genetics education—would need to be directed at this poorly informed subpopulation. The most dramatic example in this regard is cystic fibrosis: Among those at the lowest GKI levels, quite few note the disease's genetic base, while those at the very highest level are clearly aware of this circumstance. Other diseases in this category are dwarfism, color blindness, and Mongolism.
- Certain diseases, with genetic or mixed etiologies, possibly including both genetic predispositions and epidemiological and environmental positive factors, are widely viewed as linked to genetics. In these instances, differences in perception appear between men and women: Men who are more knowledgeable are more inclined to attribute genetic origins, in contrast to the less knowledgeable. In this category, we find allergy and asthma.
- For certain conditions, such as heart disease, cancer, leukemia, kidney disease, and mental illness, fairly substantial subgroups of respondents tend to suspect genetic origins. Patterns of association with GKI level are varied: There appears no trend for heart disease or cancer. For kidney disease and leukemia, the proportion assuming some genetic cause tends to rise with increasing GKI level. As noted, for mental illness opposite trends appear—increase in perception of genetic origin with increasing GKI level for males, and a somewhat puzzling trend in the opposite direction for females, with the least genetically knowledgeable women most likely to view mental illness as relating to genetic factors.
- Finally, nongenetic diseases almost universally perceived as nongenetic. Examples in this category are rheumatic fever, gonorrhea, German measles, and pneumonia.

Findings at hand show a considerable complexity in the way in which various disease types are perceived, even within a fairly homogeneous respondent population. It is also interesting to note that, *while genetic disease as such is generally considered quite rare, as shown by previous study results, genetic factors are seen at work quite broadly across a substantial range of disease types.* These include not only the specifically genetic diseases such as TSD, sickle cell anemia, and diabetes, but also, though expectedly in lower measure, in such pervasive and heterogeneous genres such as heart disease and cancer. One concludes that quite substantial consciousness of genetic forces is at work in the psychological world of the respondents. This awareness may serve to set the stage for additional interpretation of genetic disease impact and for possible compliance/cooperation in voluntary genetic disease screening programs. Conversely, in populations where such awareness is lacking or in which substantial misinformation is found, more intensive counterbalancing interpretive and educational efforts may prove to be necessary.

NOTE

1. An example of a systematic study of hospital admissions considering genetic etiology is provided by Clow et al. (1973: 160-166).

Chapter 6

TOWARD GENETIC DISEASE CONTROL: Views of Genetic Screening and Abortion

In considering issues relating to control of genetic illness, under what circumstances–particularly differential levels of genetics knowledge–are people prepared to participate voluntarily in genetic disease carrier screening programs? Who should take primary responsibility for conduct of such programs, and when in the person's life should the necessary test procedures be instituted? And if disabling conditions, as caused by genetic disease in offspring, are predicted, how are the attitudes held by the parents affected regarding selective abortion of the fetus?

While views and attitudes associated with the above questions and eventual behaviors are correlated imperfectly, people's perceptions of the issues necessarily establish the psychosocial context from which behavior affecting disease control eventually emerges. In this sense, by way of a prototype, it is useful to examine the attitudes of the specified urban Jewish population (composed primarily of married Jewish persons of reproductive age–a sample substantially representative of the potential client population of the California Tay-Sachs Dis-

TABLE 22 Readiness to take Test

	Item A. "Take Test if Disease Could be Treated" (II-16A) *Item B. "Take Test for Information Only" (II-16B)* *Item C. "Take Test, then Abort Pregnancy" (II-16C)*					
GKI:	*I lowest*	*II low*	*III middle*	*IV high*	*V highest*	*Total*
affirmative response: testing						
item A	84.0 (79/94)	85.6 (119/139)	89.5 (272/304)	93.5 (318/340)	94.5 (52/55)	90.1 (840/932)
item B	70.2 (66/94)	77.0 (107/139)	78.0 (237/304)	85.9 (292/340)	85.5 (47/55)	80.4 (749/932)
item C	84.0 (79/94)	89.9 (125/139)	92.1 (280/304)	97.1 (330/340)	96.4 (53/55)	93.0 (867/932)
normalized "three-test" mean	79.4	84.2	86.5	92.2	92.1	87.8

NOTE: Numerals outside parentheses refer to percent within GKI level responding affirmatively; numerals within parentheses refer to frequencies; e.g. 79 of 94 persons at lowest GKI level responded affirmatively to item A.

ease Prevention Program) toward various aspects of prospective control of human genetic disease.

GENETICS KNOWLEDGE AND ATTITUDES TOWARD CONTROL BY GENETIC SCREENING

Inquiry was made regarding attitudes expressed toward readiness to take an appropriate test under three distinct circumstances:

(a) if the disease could be treated;
(b) if the test were for information only (without possibility of treatment); and
(c) if the test result were to establish likelihood of prospective genetic illness, with the possibility that the pregnancy might be aborted, if subsequent tests prove the fetus to be affected with the disease.

Affirmative responses to each of these questions are shown in Table 22. It is clear that, for all testing conditions considered, readiness to take the specified test is extremely high, ranging from about 70 to 97%. In spite of this relative narrowness of range, there are material differences for the several test conditions, and for various GKI levels. The conclusions may be stated as follows:

(1) If the test serves to set the stage for a possible therapeutic abortion to end at-risk pregnancy (item c), the proportion of affirmative response, indicating readiness to take the test, generally is slightly higher than that found for the other testing conditions (items a and b, "treatment" or "information only"). This holds (though with little by way of statistical significance) at all genetic knowledge levels save the lowest.

(2) The test for information only condition elicits relatively least popular support.

(3) With respect to item c, providing a basis for assertive control, 93% of the total population express readiness to take the test, while at the other extreme, if the test is for information only (item b), the corresponding percentage is somewhat lower, near 80%.

(4) Knowledgeable respondents, at the high or highest GKI levels (IV or V), are somewhat more likely to respond affirmatively to the opportunity for testing generally than those at relatively less-informed lower levels; this is indicated by the normalized "three-test" mean computed for each GKI level, representing equal weights for each of the testing conditions, and by comparison of affirmative responses among GKI levels for each item. Persons in the least knowledgeable (lowest) GKI category respond affirmatively to testing (as indicated by the normalized three-test mean) at 79%, while those most knowledgeable, in the high and highest GKI categories, respond affirmatively at about 92%.

GKI AND CONTROL

Major Institutional/Individual Roles in Testing

The inquiry addresses the issue—in connection with both diagnostic testing and risk testing—of the preferred loci of

TABLE 23 Major Role in Risk Test

Item A. "State Government – by Law" *(IV-1A)*
Item B. "Community (potentially affected religious or ethnic group)" *(IV-1B)*
Item C. "Medical Profession" *(IV-1C)*
Item D. "The Individual (independent of outside influences) *(IV-1D)*

GKI:	*I* lowest	*II* low	*III* middle	*IV* high	*V* highest	*Total*
item A	37.2 (35/94)	38.8 (54/139)	44.1 (134/304)	43.5 (148/340)	52.7 (29/55)	42.9 (400/932)
item B	25.5 (24/94)	25.2 (35/139)	28.3 (86/304)	26.2 (89/340)	25.5 (14/55)	26.6 (248/932)
item C	72.3 (68/94)	61.9 (86/139)	68.4 (208/304)	66.8 (227/340)	67.3 (37/55)	67.2 (626/932)
item D	64.9 (61/94)	75.5 (105/139)	71.7 (218/304)	71.8 (244/340)	65.5 (36/55)	71.2 (664/932)
Individual/ Government ratio (D./A.)	1.74	1.95	1.63	1.65	1.24	1.66
Medical/ Government ratio (C./A.)	1.94	1.60	1.55	1.54	1.28	1.57
Individual/ Medical ratio (D/C)	.90	1.22	1.05	1.07	.97	1.06

responsibility with regard to implementation of the testing program. The following alternative loci are specified:

(a) state government, by law;
(b) the community (the potentially affected religious or ethnic group);
(c) the medical profession; and
(d) the individual (independent of outside influences).

Important consistencies appear in findings relating to major roles preferred in both fetal diagnostic and risk testing to determine carrier-status; the following are salient findings (see Table 23).

(1) For the sample as a whole (composed of generally well-educated persons), individual responsibility is preferred to any of the

alternative responsibility loci; more than 70% choose this mode of responsibility assignment in diagnostic and in risk testing (GKI level seems to make little difference in this regard).

(2) Following preference for individual responsibility, generally by a narrow margin, responsibility assignment to the medical profession is proposed.

(3) Quite far behind endorsement of individual and medical responsibility follow "state government" and "community," that is, the potentially affected religious or ethnic group itself. Though state government does not rank high on a relative basis, it is of interest to note that some measure of governmental intervention in this personally sensitive area is approved by substantial absolute proportions of respondents (particularly in risk testing), with approving percentages ranging from about 25 to 53%, depending on conditions proposed.

In further consideration of possible association between genetics knowledge and views concerning responsibility for control procedures, and for purpose of other comparison, one may construct a simple measure to show the extent to which any one responsibility choice is more or less preferred over another; for instance, note the individual/government ratio (D/A), in Tables 23 and 24. This measure is derived by dividing the proportion preferring individual initiative by the corresponding proportion preferring state government responsibility. It is found consistently that the ratio showing relative preference for individual responsibility is greater in the case of diagnostic testing than it is in the case of risk-testing: *Individual responsibility is clearly preferred in diagnostic testing.* Particularly in risk testing, there is some tendency for respondents in the lower GKI categories (I, II) to prefer individual responsibility over government responsibility still more so than those at higher GKI levels. *The most knowledgeable respondents, though still stressing the role of the individual above other alternatives, appear somewhat more ready to accept government participation* in establishment of testing programs than are the less knowledgeable (Table 23, D/A to GKI I = 1.74; to GKI V = 1.24).

TABLE 24 Major Role in Diagnostic Test

Item A. "State Government by Law" (IV-2A)
Item B. "Community (potentially affected religious or ethnic group)" (IV-2B)
Item C. "Medical Profession" (IV-2C)
Item D. "The Individual (independent of outside influences) (IV-2D)

GKI:	*I lowest*	*II low*	*III middle*	*IV high*	*V highest*	*Total*
item A	30.0 (28/94)	25.2 (35/139)	29.9 (91/304)	32.6 (111/340)	36.4 (20/55)	36.4 (285/932)
item B	20.2 (19/94)	15.1 (21/139)	21.1 (64/304)	20.9 (71/340)	21.8 (12/55)	20.1 (187/932)
item C	74.5 (69/94)	65.5 (91/139)	69.7 (212/304)	70.3 (239/340)	70.9 (39/55)	69.8 (651/932)
item D	73.4 (69/94)	78.4 (109/139)	77.0 (234/304)	75.3 (256/340)	76.4 (42/55)	76.2 (710/932)
individual/ Government ratio (D./A.)	2.45	3.11	2.57	2.31	2.10	2.09
Medical/ Government ratio (C./A.)	2.48	2.60	2.33	2.16	1.95	1.92
Individual/ Medical ratio (D/C)	.99	1.20	1.10	1.07	1.08	1.09

An analogous ratio is constructed with reference to preference for responsibility by the medical profession as contrasted with intervention by state government; this is the medical/government ratio (C/A). *Greater readiness to grant responsibility to the medical profession in both risk and diagnostic testing is found at the lower GKI levels.* Further, this tendency is more distinct for diagnostic than for risk-testing. Generally, *less knowledgeable persons place greater faith in medical responsibility than do more knowledgeable respondents,* when comparison is made with the alternate option of government involvement.

Finally, a ratio is constructed to contrast relative assignment of responsibility for individual and medical profession, the individual/medical ratio (D/C). There are no substantial differences in this ratio among GKI levels, with the ratios hovering about a relatively even split (1.0) though, in accordance with findings reported above, individual responsibility

TABLE 25 "When Should Couple Take Test?"

GKI	I lowest	II low	III middle	IV high	V highest	Total
no prevention						
before engagement	10.6	9.4	10.5	9.1	14.5	10.1
	(10)	(13)	(32)	(31)	(8)	(94)
before marriage	42.6	43.9	44.4	50.0	40.0	45.9
	(40)	(61)	(135)	(170)	(22)	(428)
before family	24.5	30.9	31.3	26.8	20.0	28.2
	(23)	(43)	(95)	(91)	(11)	(263)
during pregnancy	4.3	15.5	9.5	8.8	14.5	9.0
	(4)	(13)	(29)	(30)	(8)	(84)
other	2.1	2.2	1.3	2.6	3.6	2.1
	(2)	(3)	(4)	(9)	(2)	(20)
D.K., N.A.	16.0	4.3	3.0	2.6	7.3	4.6
	(15)	(6)	(9)	(9)	(4)	(43)
	100.0[a]	106.2[b]	100.0	100.0	100.0	100.0
with prevention						
before engagement	7.4	4.3	9.5	6.5	9.1	7.4
	(7)	(6)	(29)	(22)	(5)	(69)
before marriage	31.9	33.8	32.6	33.8	29.1	32.9
	(30)	(47)	(99)	(115)	(16)	(307)
before family	36.2	38.1	34.9	31.8	32.7	34.2
	(34)	(53)	(106)	(108)	(18)	(319)
during pregnancy	11.7	17.3	18.1	22.4	23.6	19.2
	(11)	(24)	(55)	(76)	(13)	(179)
other	2.1	2.2	0	2.1	1.8	1.4
	(2)	(3)	(0)	(7)	(1)	(13)
D.K., N.A.	10.6	4.3	4.9	3.5	3.6	4.8
	(10)	(6)	(15)	(12)	(2)	(45)
	100.0	100.0	100.0	100.0	100.0	100.0

a. Percentages may not add to exactly 100 due to rounding.
b. Total exceeds 100.0 due to multiple answer (response error).

continues to be preferred slightly over responsibility by the medical profession.

Conditions Concerning Timing of Test

Table 25 shows the preferred timing, under conditions of "no prevention"—namely, when an adult obtains results of a blood test, indicating carrier status, without other recourse;

and "with prevention"—that is, in a fetal test with the option of therapeutic abortion available. The following are the principal results:

(1) Under condition of no prevention, the time before marriage (nearly 46%) is generally preferred. In order, the next choices are before family (28%), before engagement (10%), and during pregnancy (9%).
(2) Under conditions with prevention, the pattern shifts slightly. Now it is before family that receives the highest percentage of choice (34%), followed by before marriage (33%). During pregnancy is in third position (19%), rather than before engagement (7%).
(3) As to comparisons among GKI levels, the results are generally not significant; within the demographically homogeneous sample population, genetics knowledge seems to make little difference in timing preferred for administration of a specified test.

Attitudes Toward Abortion

Attitudes toward abortion—often closely linked to ideological and religious value positions—have been examined in numerous prior inquiries. Such attitudes, as considered here, are relevant to the extent that they may establish a basis for the carrier couple to choose remedial intervention, such as selective abortion, if potential genetic disease in the fetus is indicated by a specified testing procedure.

The data show (see Table 26) that *a significant majority of the Jewish, well-educated respondents (68%) believes that abortion should be available on request.* An additional 30% favor abortion sometimes, under particular circumstances. Thus, *the proportion totally against abortion is minimal.*

These findings are generally consistent with data showing the "liberal" ideological and value system found to characterize Jewish populations. While Orthodox Jews may represent more traditional positions, their inclusion in the present

TABLE 26 Jewish Religious Ideology and Attitude toward Abortion

	Orthodox	*Conservative*	*Reform*	*Nonpractic.*	*Other*	*Total*
male						
on request	(5) 50.0	(93) 62.0	(77) 67.5	(68) 77.3	(13) 65.0	(257) 66.9
totally against	(0) 0	(2) 1.3	(1) 0.9	(1) 1.1	(2) 10.0	(6) 1.6
sometimes	(5) 50.0	(54) 36.0	(36) 31.6	(19) 21.6	(5) 25.0	(120) 31.3
other	(0) 0	(1) 0.7	(0) 0	(0) 0	(0) 0	(1) 0.3
	(10)100.0	(150)100.0	(114)100.0	(88)100.0	(20)100.0	(384)100.0
female						
on request	(4) 50.0	(118) 70.7	(118) 74.2	(44) 84.6	(15) 46.9	(300) 71.6
totally against	(0) 0	(0) 0	(0) 0	(0) 0	(5) 15.6	(5) 1.2
sometimes	(4) 50.0	(48) 28.7	(40) 25.2	(8) 15.4	(12) 37.5	(112) 26.7
other	(0) 0	(1) 0.6	(1) 0.6	(0) 0	(0) 0	(2) 0.4
	(8)100.0	(167)100.0	(159)100.0	(52)100.0	(32)100.0	(419)100.0
					n = 803	

NOTE: Non-Jewish respondents not included in this table.

sample is of minor proportion, expected as Orthodox constitute only a small proportion of the younger group.

The relationship between Jewish religious ideology and attitude toward abortion is shown (for *Jewish* married respondents only), in Table 26. Note that less than 3% of respondents identify as Orthodox, nearly 40% as Conservative, 34% as Reform, and 24% as nonpracticing or as associated with some other ideological position. While the number of Orthodox is too small to permit rigorous analysis, the general relationship linking Orthodox views and more cautious, qualified attitudes toward abortion is generally supported. Consistently, *as the continuum ranges from Orthodoxy to the nonpracticing, the proportion approving abortion on request increases.* Taking account of the ideological pattern generally skewed toward the liberal end of the continuum, it is evident that, for each ideological group, at least one-half, and often 70 or 80% favor abortion on request. As expected (with exception of the numerically limited Orthodox category), higher proportions of females endorse abor-

TABLE 27 Conditions Under Which Sometimes Favor Abortion

	Male (*n* = 120) = 100%		*Female* (*n* = 112) = 100%	
mother's physical health	(118)	98.3	(106)	94.6
mother's mental health	(112)	93.3	(98)	87.5
family too large	(57)	47.5	(62)	55.4
teenage mother	(22)	18.3	(47)	42.0
unmarried mother	(26)	21.7	(44)	39.3
mother over 40	(22)	18.3	(45)	40.2
other spec. case	(32)	26.7	(41)	36.7

TABLE 28 Attitudes Toward Abortion

GKI:	*I*	*II*	*III*	*IV*	*V*	(*III-1*) *Total*
on request	50.0 (47)	61.9 (86)	69.1 (210)	74.7 (254)	69.1 (38)	68.1 (635)
sometimes favor	45.7 (43)	36.7 (51)	28.9 (88)	24.1 (82)	27.3 (15)	29.9 (279)
totally against	0 (0)	0.7 (1)	1.6 (5)	0.9 (3)	3.6 (2)	1.2 (11)
N.A., D.K.	4.3 (4)	0.7 (1)	0.3 (1)	0.3 (1)	0 (0)	0.7 (7)
	100.0	100.0	100.0	100.0	100.0	100.0 (932)

NOTE: Respondents over age 49 included in this table.

tion without qualification than do males in the corresponding category.

Considering only those who favor abortion under specified conditions, it is found that abortion is widely—nearly universally—approved (87 to 98%) if the mother's physical or mental health may be endangered (see Table 27). About one-half would favor abortion if the family were regarded as too large.

For conditions as above the differences in attitude between males and females are slight. However, for other conditions, *important sex-linked contrasts appear,* with, in each

TABLE 29 Attitude Toward Abortion Under Specified Conditions and Age of Predicted Death

	yes	*no*	*DK/NA*	*Total*
Slight physical handicap				
male	(91) 23.7	(198) 51.6	(95) 24.7	100.0%
female	(102) 24.3	(177) 42.2	(140) 33.4	100.0
Severe physical handicap				
male	(311) 81.0	(12) 3.1	(61) 16.0	100.0
female	(355) 84.7	(10) 2.4	(54) 12.9	100.0
Mild mental retardation				
male	(193) 50.3	(93) 24.2	(98) 25.5	100.0
female	(226) 53.9	(75) 17.9	(118) 28.1	100.0
Severe mental retardation				
male	(346) 90.1	(6) 1.6	(32) 8.4	100.0
female	(376) 89.7	(5) 1.2	(38) 9.1	100.0
Death before 5				
male	(336) 87.5	(11) 2.9	(37) 9.7	100.0
female	(376) 89.7	(8) 1.9	(35) 8.4	100.0
Death before 10[a]				
male	(298) 77.6	(22) 5.7	(64) 16.7	100.0
female	(338) 80.7	(14) 3.3	(67) 16.0	100.0
Death before 20				
male	(250) 65.1	(44) 11.5	(90) 23.5	100.0
female	(283) 67.5	(27) 6.4	(109) 26.0	100.0

Totals shown horizontally.

a. The question wording as death before 10 not entirely precise in the context of the prior question; death before 5 might have been more accurately phrased as "death between ages 5 and 10." However, as indicated in survey pretests, the chosen wording is interpreted much along the lines as death between ages 5 and 10, while avoiding somewhat confusing repetition of the numerals 5 and 10. A similar condition holds for the item death before 20, with reference to the two questions preceding, as connoting "death between (ages) 10 and 20."

instance, the female more prepared to endorse abortion than the male. In this respect, *females are more favorably inclined toward abortion if the pregnancy occurs in a teenager, if the potential mother is unmarried, or if the potential mother is over forty.*

Those at the lowest GKI level are somethat less likely to express approval of abortion on request than persons with relatively more knowledge of genetic disease (see Table 28).

Finally, we wish to inquire whether the respondents would favor abortion if it could be determined early in pregnancy

(3-4 months) that the child resulting from that pregnancy would *definitely* suffer from a specified set of conditions, or be subject to a disease that would lead to death at a specified age. Table 29 summarizes relevant responses. Generally in accordance with previously noted readiness to approve of abortion when serious issues of physical and mental health are involved, abortion is readily approved (by 80% or more of the respondents) if a severe physical handicap or severe mental retardation of the child is diagnosed early in pregnancy. *Severe mental retardation is regarded as cause for abortion by a slightly higher proportion than if a severe physical handicap is forecast.* Similarly, mental retardation is viewed as a more critical issue than physical handicap at lesser levels of disability as well; About 50% approve abortion in cases of mild mental retardation, as compared to about 24% approving if a slight physical handicap is likely to affect the newborn child.

Nearly 90% approve of abortion if death of the child is predicted at five years of age or younger. The percentages endorsing abortion drop somewhat, to levels around 80%, if death is expected before age ten, and to approximately 60% if death is expected before age twenty. (In these and similar comparisons, differences in viewpoint between the sexes are negligible.)

DISCUSSION

In the urban, primarily Jewish, married population studied, expressed readiness to take tests as a basis for potential control of genetic disease, or simply "for information" regarding possible genetic disease in offspring, is very high indeed. It is, however, *relatively* lowest for persons with the least amount of genetics knowledge, and under conditions when the test is for information only, without possibility of therapeutic intervention or prevention. *The high points of the range of readiness to seek testing, above 96%, are found for the most highly informed (GKI levels IV, V), if the test may serve to indicate a condition for which prevention can be achieved through abortion.*

These very considerable verbal affirmations of "test readiness" would seem to indicate the existence of a favorable *attitude* climate as a basis for implementation of appropriate testing programs. It must be recalled, however, that such implementation of large-scale genetic disease testing depends not only on attitudes expressed by a particular clientele; rather, effective interpretation of the program's utility by means of carefully designed and effectively conducted education programs is required to assure compliance by a substantial segment of the potential client population. By way of analogy—and in a sense, substantively—attitudes overtly favorable toward genetic testing need to be transformed into action with the aid of marketing approaches, such as for example, advertising, mass merchandising, and purposeful public relations procedures.

While the client is subject to persuasion by means noted, the issue remains as to who should be responsible for determining whether or not a person is to take the specified genetic test. As indicated, the principal responsibility is generally assigned to the individual, ostensibly independent of outside influence. Interpretation of this response must recognize, of course, that while such autonomy in decision-making may represent a personal value held dearly by the respondents, sociologically the person typically *is* subject to a variety of "outside" influences, such as education, information, and medical suggestions that ultimately act with value predispositions to determine compliance or noncompliance.

While, with minor differences, particularly at the lowest genetics knowledge level, the assertedly independent individual is seen as the preferred decision locus, and, while medical intervention is valued widely, without exception it is the community of the potentially affected religious or ethnic group that comes in last among the four specified decision alternatives. This holds both for the risk test and for the diagnostic test; only 20 and 25% of respondents believe that the (in this case Jewish) community should be primarily responsible for testing, while, as reported, 65 to nearly 80% believe that it is the autonomous individual who should decide. It is clear that the prevalent Jewish population attitude places little credibility in the community as a potential decision-making force in either risk-testing or diagnostic-testing as genetic disease control procedures, in spite of a common assumption regarding the cohesive social impact of Jewish community institutions.

A further key aspect of genetic disease control is that of the *timing* of the specified test. It is reported that, under conditions of no prevention, the time *before* marriage is preferred by a plurality of the respondents. This preference may be associated with the view that if the illness cannot be prevented by any means, such testing might yield information which potentially might dissuade a couple from marriage. The next preferred option, testing before family is started, would suggest that minimally the couple might be influenced by the information elicited by the test in their plans for or avoidance of conception.

With prevention, however, less concern appears as to the possible implications of the test result for marriage as such. Now 33%, a smaller proportion than was the case under no prevention conditions (46%), believes that the test should be implemented prior to marriage. The fact that meaningful differences in preferred timing appear suggests that respondents are capable of discerning with promising understanding the implications of genetic testing for marriage foundation and family planning.

While anticipated differences in liberality regarding abortion appear among the several ideological categories in the population, the clear-cut endorsement of "abortion on request" stands out in the population *as a whole:* Two-thirds or more hold that abortion should be available on request. Females exceed males in this respect by a slight margin only

(72 versus 67%; see Table 26). It is evident that the "attitude toward abortion" climate prevalent in the Jewish population establishes a ready basis for implementation of therapeutic abortion if genetic illness is diagnosed and predicted for the unborn child.

Particularly with the rather small proportion of Orthodox, the subgroup showing relatively the highest proportion of reluctance regarding on-request abortion, it is evident that *the reproductive-age Jewish population as a whole is well prepared psychologically to accept and implement medically indicated abortion procedures.*

Even for those who fail to favor abortion on request, choosing rather to specify conditions under which abortion may be (sometimes) acceptable, a distinctly liberal attitude base emerges. Among those who are willing to accept abortion conditionally, about nine of ten accept abortion as appropriate if either the mother's physical mental health is endangered. Here, perhaps unexpectedly, differences between the sexes are minor. However, for other conditions affecting the desirability of abortion (teenage mother, mother over 40, and the like), females, more so than males, are prepared to accept abortion as a remedy. Sex-linked contrasts are greatest when abortion is proposed because the mother is either a teenager or over 40, and somewhat less distinct if the mother is unmarried or if the family regarded as too large.

By way of net effect, the conditional responses further support the contention that even in those instances in which abortion on request is not directly approved, the possibility of abortion is held open, under specified conditions. Indeed, one may wonder about something of a fairly widespread "intolerance for imperfection," with about one-half of the respondents opting for abortion if only *mild* mental retardation is expected; and near one-fourth expressing a pro-abortion view if only a *slight* physical handicap is forecast for the newborn child.

With regard to possible damage in the offspring, severe health problems for the child (mental more than physical) are, of course, viewed as compelling reasons for abortion. The attitude that *mental* retardation is relatively a more serious obstacle to satisfactory living than a *physical* handicap is noteworthy. A concern with intellectual or mental life, with the importance of learning, and with similar value positions frequently attributed to Jewish populations—particularly those who are young and highly educated—appears reflected in these judgments.

With reference to the prospects for a relatively early (before age five) death of the child, the vast majority of respondents (about nine in ten) are prepared to recommend abortion. While corresponding percentages decline to about eight in ten if death may occur before age ten and to approximately two of three if death is predicted before age twenty,[1] *it is evident that any forecast of death in childhood or early adulthood constitutes a sufficient cause for abortion* in the view of most persons in the study population. While other important religions and ideologies often are heavily concerned with the significance of "life at conception" or "life in utero," the Jewish value position appears to place primary emphasis on "life in living," particularly beyond the formative stages of growth and into the years of full maturity.

SUMMARY

The inquiry reported suggests the existence of a firm attitude base in the subject population (principally Jewish, married, of reproductive age, in Los Angeles) for autonomous readiness to participate in appropriate genetic testing programs—voluntarily, with medical support, rather than at behest of government or community—and for seeking therapeutic abortion when medically indicated, on request or in a wide variety of specified circumstances. Even among the Orthodox, a small minority within the Jewish population, abortion appears widely acceptable if conditions warrant, though the present data base is too small to permit definitive statements. Among the Conservative and Reform—the more numerous groups in the Jewish population—acceptance of abortion is widespread. Consistently, physical and mental health, particularly the latter, constitute a primary consideration in endorsement of abortion, perhaps to a point of possible intolerance for imperfection. Conditions of life and living, rather than the very act of conception, appear of primary concern to the population studied; the attitude pattern, stressing individual responsibility and pervasive acceptance of therapeutic abortion, establishes a substantial foundation for voluntary genetic disease carrier screening and similar human genetic disease control efforts.

NOTE

1. As noted earlier, modified question wording specifying age *ranges* of expected death—between ages 5 and 10; or between ages 10 and 20, as compared with wording as actually employed in questionnaire has little substantive significance.

Chapter 7

BEING A CARRIER: IS THERE STIGMA?

Does a person who is identified as a genetic disease carrier by means of a blood test screening program feel stigmatized? While this question has, from time to time, been asserted to warrant an affirmative reply, the relevant available data base has been excessively limited, often in terms of geography and cultural specificity. Perhaps more serious, however, is the lack of conceptual clarity concerning the meaning of stigma.

It is essential to distinguish between stigma as derived from stigmata, the classic psychopathic symbols of the wounds of Christ, or similar deeply incisive and damaging imprints of a physiologic or psychologic nature, and stigma as a transient state of discomfort or concern. The results presented here indicate that, at least as relates to Tay-Sachs disease, some discomfort or concern surely is found among carriers, but there is no sense of stigma in the classic sense of the term. Other data suggest as well the positive context in which the knowledge of carrier status is received, with continuing endorsement of the blood test screening concept. Still, tensions of various kinds arise as carrier status becomes known in the framework of the family.

A detailed personal-interview follow-up study was conducted, including 74 carriers of the Tay-Sachs gene, in the greater Los Angeles area (1976-1977). The very few carrier couples, directly at risk, were excluded from the inquiry. Spouses of the carriers also were interviewed. This feature makes possible a comparison of attitudes within the marital dyad and serves as an internal control. The carrier/spouse interviews were conducted quite independently of one another.[1] From the substantial structured and open-end interview protocols, selected major findings are highlighted here.

Table 30 reports the demographic characteristics of TSD carriers responding in this study phase; an almost perfect response rate, 98.7%, was attained, exclusive of 11 carriers who had moved out of state or for whom no current address was available, in spite of concerted efforts to locate.

Among the 74 carriers successfully contacted (not considering their spouses, an additional 74 interviews, for a total of 148), about three-fourths are in the 30 to 44 age range, mostly in the early and mid-thirties. Females outnumber males by a small margin.

As is typical for virtually all U.S. Jewish population samples of comparable age, the education level is exceptionally high: some 7 of every 10 respondent carriers have at least completed college or have been granted an advanced or professional degree.

In Jewish religious ideology, the respondents follow a predominantly mainstream pattern. The majority identify, regardless of formal congregation membership, as Reform or Conservative, with a slight and somewhat atypical preponderance of Reform. The proportion of Orthodox is small, of an order of magnitude similar to that found in the Jewish population as a whole. The configuration of these significant variables suggests that the carrier sample studied is quite comparable to broader Jewish population samples of reproductive age.

What about the carrier's response upon first being informed of the positive result of the screening blood test? And

TABLE 30 Demographic Characteristics–TSD Carriers

	age/sex				
males	29, under[a]	(7)	9.5	total males:	46.0
males	30, up[b]	(27)	36.5		
females	29, under	(13)	17.5	total females:	54.0
females	30, up	(27)	36.5		
		(74)	100		
	education				
high school or less		(8)	10.8		
some college		(15)	20.3		
college graduate		(29)	39.2		
advanced or professional degree		(22)	29.7		
		(74)	100		
	Jewish religious ideology/identification				
Orthodox		(5)	6.8		
Conservative		(22)	29.7		
Reform		(26)	35.1		
"just Jewish"		(19)	25.7		
other		(2)	2.7		
		(74)	100		

a. primarily 20-29
b. primarily 30-39

what about the response of his/her spouse? The findings appear in Table 31. The resulting picture is striking: There is indeed substantial *initial* concern by the carrier regarding his/her circumstance, when the news first becomes known. It must be noted that, at this early stage in the person's experience as carrier, the implications of carrier status—including its absence of negative significance for the person's health—typically are not clearly known or fully appreciated by the carrier. Even in this context only about one carrier in five

TABLE 31 "When First Found Out That TSD Carrier"

		A *deeply troubled*	*B* *quite concerned*	*C* *only mildly concerned*	*D* *quite un-concerned*	*Total*
Carrier re self	(CA-12)	(16) 21.6	(22) 29.7	(26) 35.1	(10) 13.5	(74) 100
Spouse re carrier	(SP-6)	(7) 9.5	(32) 43.2	(27) 36.5	(8) 10.8	(74) 100
Carrier re spouse	(CA-13)	(5) 6.8	(15) 20.3	(36) 48.6	(18) 24.3	(74) 100
Spouse re self	(SP-7)	(5) 6.8	(18) 24.3	(36) 48.6	(15) 20.3	(74) 100

(CA-12) Thinking back once more to the day *when you first found out* that you were a TSD carrier . . . which of the following statements *then* best reflected your reactions?

(SP-6) Thinking back for a moment to the day your spouse *first found out* that he/she was a TSD carrier . . . which of the following statements *then* best reflected his/her reactions?

(CA-13) Which of these statements best describes how your spouse felt *then* . . . when he/she first found out that you were a TSD carrier?

(SP-7) Which of these statements best describes how *you* felt *then* . . . when he/she first found out that you were a TSD carrier?

regards himself/herself as "deeply troubled" concerning the matter. An additional 30% characterize their feeling as "quite concerned." These proportions are roughly comparable to the proportion of carriers reporting themselves as "upset," in accordance with a prior study by Childs et al. (1976). As will be shown later, the intensity of concern, or worry, eases further as time elapses, and as TSD information bulletins and other means clarify matter of health impact, reproductive risk, and the like.

One notes a pattern to be examined shortly in further detail: The carrier's spouse, while acknowledging the carrier's sense of concern, views his/her condition as somewhat *less* severe than the carrier himself/herself. While nearly 22% of carriers indicated that they were "deeply troubled" when first finding out about their carrier status (with little information at this time as to the implications), only some 10% of spouses regard their partner/carrier as troubled at this level of intensity. In a sense, the spouses show a net tendency to somewhat minimize the severity of the carriers' responses, though generally acknowledging that something anxiety-provoking had occurred.

As might be expected, the response of the carrier's *spouse* regarding his/her own current status is considerably less severe than that reported by the carrier. The typical reaction is one of mild concern, though about equal numbers describe their feelings as quite concerned or quite unconcerned.

Tables 32 and 33 contrast the carrier's feelings, and the spouse's feelings, respectively. As the interviews were conducted independently, with no opportunity for carrier and spouse to discuss results or procedure, a systematic comparison is possible between the carrier's self-perception and the spouse's perception of the carrier's response. In 31% of instances, both members of the couple agreed as to the level of concern shown by the carrier. However, in the typical instance, particularly if the carrier's response tended to be a sense of being deeply troubled or quite concerned, the spouse's assessment was more optimistic than the carrier's self-perception—that is, when matters were seen by the carrier "through a glass darkly," the spouse was somewhat more inclined to perceive his/her mate's reaction as a bit brighter or less severe. Conversely, when the carrier described himself/herself as only mildly concerned, the partner's response tended to be somewhat more pessimistic. At any rate, the intrusion of anxiety seemed, somehow, to be less conducive to accurate perception of feelings within the marriage dyad than under less-tense circumstances: The proportion of "accurate" judgments of feeling was considerably higher when the *spouse's* feelings, rather than the carrier's, were assessed. Here the batting average of correct judgments proved to be more than 55%, as compared with the 31% when the carrier's feelings were assessed.

TABLE 32 Comparison: "When first found out that carrier": The Carrier's Feelings Re Self, and Spouse's Perception of Carrier's Feelings

Carrier's feelings re self (CA-12) / Spouse's perception of carrier's feelings (SP-6)		*A* *deeply troubled*	*B* *quite concerned*	*C* *only mildly concerned*	*D* *quite un-concerned*	*Total*
deeply troubled	A	(3) 4.1	(4) 5.4	(-) 0	(-) 0	(7) 9.5
quite concerned	B	(7) 9.5	(10) 13.5	(14) 18.9	(1) 1.4	(32) 43.2
mildly concerned	C	(6) 8.1	(8) 10.8	(7) 9.5	(6) 8.1	(27) 36.5
quite unconcerned	D	(-) 0	(-) 0	(5) 6.8	(3) 4.1	(8) 10.8
Total		(16) 21.6	(22) 29.7	(26) 35.1	(10) 13.6	(74) 100
Spouse's perception						
more optimistic		17.6	10.8	6.8	x	35.2
more pessimistic		x	5.4	18.9	9.5	33.8
accurate		4.1	13.5	9.5	4.1	31.2

NOTE: Numerical discrepancies due to rounding.

How does the determination that one or the other partner is a TSD carrier affect the couple's plans and relationships? Table 34 summarizes plans to have additional children, as seen by the carrier and by the spouse. Age differences are highly significant, and some sex differences appear: Not surprisingly, those 29 years old or younger, both males and females, are more likely to plan additional children than are those 30 years old and older. Female spouses of carriers are less inclined to want additional children than are male spouses. One may speculate that, though wide agreement is asserted regarding plans to have children (see Table 34), the female *spouse* may feel more personally threatened, as a possible self-defined victim of circumstance (whether or not she might use the term), than the female *carrier* who in some

TABLE 33 Comparison: "When first found out that carrier": Carrier's Perception of Spouse's Feelings, and Spouse's Feelings Re Self

Carrier's perception of spouse's feelings (CA-13)		*A* *deeply troubled*	*B* *quite concerned*	*C* *only mildly concerned*	*D* *quite unconcerned*	*Total*
Spouse's feelings re self (SP-7): deeply troubled	A	(2) 2.7	(2) 2.7	(1) 1.4		(5) 6.8
quite concerned	B	(1) 1.4	(6) 8.1	(9) 12.2	(2) 2.7	(18) 24.3
mildly concerned	C	(2) 2.7	(7) 9.5	(22) 29.7	(5) 6.8	(36) 48.6
quite unconcerned	D	(-) 0	(-) 0	(4) 5.4	(11) 14.9	(15) 20.3
Total		(5) 6.8	(15) 20.3	(36) 48.6	(18) 24.3	(74) 100
Carrier's perception						
more optimistic		4.1	14.9	6.8	x	25.8
more pessimistic		x	1.4	12.2	5.4	19.0
accurate		2.7	8.1	29.7	14.9	55.4

NOTE: Numerical discrepancies due to rounding.

sense, explicitly or implicitly, takes responsibility for the ensuing events, particularly if she is at the very height of her childbearing potential—that is, 29 or under.

Least likely to want additional children are female spouses, 30 years old and older (10%), while most likely to want children are male spouses 29 and under (89%), and female carriers 29 and under (85%).

Table 35 shows the reported agreement/disagreement between husbands and wives regarding plans to have additional children. Wide agreement is indicated with reference to plans for additional births, nearly 90% by both spouses. In a majority of cases, between 50 and 60%, no additional children are planned.

TABLE 34 Plans To Have Additional Children?

Carrier (CA-19)	*M* *29, under*	*M* *30, up*	*F* *29, under*	*F* *30, up*	*Total*
yes, plans additional children	(5) 71.4	(7) 25.9	(11) 84.6	(4) 14.8	(27) 36.5
no, does not plan additional children	(1) 14.3	(20) 74.0	(2) 15.4	(22) 81.5	(45) 60.8
don't know	(1) 14.3	(-) 0	(-) 0	(1) 3.7	(2) 2.7
	(7) 100	(27) 100	(13) 100	(27) 100	(74) 100
Spouse (SP-12)					
yes, plans additional children	(8) 88.9	(8) 26.7	(8) 57.1	(2) 9.5	(26) 35.1
no, does not plan additional children	(1) 11.1	(20) 66.7	(2) 14.3	(19) 90.5	(42) 56.8
don't know	(-) 0	(2) 6.7	(4) 28.6	(-) 0	(6) 8.1
	(9) 100	(30) 100	(14) 100	(21) 100	(74) 100

NOTE: Considered as couples, rather than as carrier/spouse separately, those with two or more children were less likely to plan additional children than those with one child, or the childless.

Given intentions as noted, both carriers and their spouses claim, in upward of 90% of the cases, that their plans to have or not have children were not affected at all by the determination of carrier status. Likewise, feelings regarding birth control, frequencies of sexual activity, and the seeking of medical attention also seem to be unaffected by carrier status (see Tables 36 and 37).

In spite of these denials, the context of concern previously summarized and some separately noted differences in desire for additional children suggest that forces somewhat below the surface of public assertion are at work in the couple's interpersonal dynamics when it is established that one or the other member is a TSD carrier.

The issue of stigma may be approached in still another way. The circumstances reported, though indicative of some

TABLE 35 Husband/Wife Agreement/Disagreement Re Plans To Have Additional Children

	Carrier's perception (CA-21)		*Spouse's perception (SP-12)*	
yes, plans additional children	(28)	37.8	(26)	35.1
no, does not plan additional children	(44)	59.5	(42)	56.8
don't know	(2)	2.7	(6)	8.1
	(74)	100	(74)	100

Carrier's view re agreement/disagreement with spouse re plans to have children (CA-21-B):

agree	(65)	87.8
disagree	(9)	12.2
	(74)	100

Spouse's view re agreement/disagreement with carrier re plans to have children (SP-14)

agree	(70)	94.6
disagree	(3)	4.1
not discussed	(1)	1.4
	(74)	100

disequilibrium for carrier and spouse, do not suggest that stigma, in traditional use of the term, is descriptive of the carrier's self-definition, or that of his/her spouse's. Both pilot studies and the data themselves suggest that a direct question to carrier or spouse of whether he/she feels stigmatized by the unfolding events would yield uniformly negative responses. Therefore, a less personally threatening question was formulated concerning the possible stigma (using that specific

TABLE 36 "Plans Re Having Children Affected by Carrier Status?"

	Carrier's view (CA-20)		*Carrier's perception of spouse's view (CA-22)*	
not affected at all	(68)	91.9	(71)	95.9
affected somewhat	(3)	4.1	(2)	2.7
affected importantly	(1)	1.4	(-)	-
don't know/other	(2)	2.7	(1)	1.4
	(74)	100	(74)	100

Spouse's view (SP-13): 100% "not affected at all"

TABLE 37 Possible Other Impact of Carrier Status

changed feelings about birth control	(CA-47-A)	yes:	(2)	2.7
changed frequency of sexual activity	(CA-48)	yes:	(0)	0[a]
seeking of medical attention	(CA-54)	more often or less often:	(0)	0

a. similar finding for spouse (SP-30)

term) that might derive for Jewish people generally as a result of TSD publicity. Results are shown in Table 38.

In less than 10% of the cases, with only negligible differences between carriers and their spouses, is publicity regarding Tay-Sachs disease seen as a kind of stigma on the Jewish people. The vast majority, about 80%, clearly view such publicity as not a stigma, agreeing that "it's okay to have such publicity."

Would carriers advise others, generally in the 20- to 49-year-old range, to take a TSD screening blood test? As is shown in Table 39, such positive advice is almost universally endorsed, some 87% suggesting that such persons *definitely* should have such a blood test "under all circumstances,

TABLE 38 Is TSD Publicity a Stigma on the Jewish People?

	Carrier *(CA-38)*		*Spouse* *(SP-24)*	
is a kind of stigma	(7)	9.5	(6)	8.2
is not a stigma	(62)	83.8	(57)	77.0
never thought about it	(5)	6.8	(7)	9.5
no opinion/other	(0)	0	(4)	5.4
	(74)	100	(74)	100

TABLE 39 Carrier's Advice to 29-49 Year Old Re Desirability of TSD Screening Blood Test (CA-41)

stay away	(1)	1.4
consider having test/ but doubts	(0)	0
probably have test/ if convenient	(9)	12.2
definitely have test/ convienient or not	(64)	86.5
	(74)	100

convenient or not." A genetics screening program for Tay-Sachs disease is seen as "absolutely essential" by more than 80% of the carriers (see Table 40). While there is practically no opposition to this kind of program (far from being turned off, carriers are unequivocal supporters of the screening concept), most believe that an effort of this nature should be voluntary rather than compulsory (see Table 41).

Is the implementation of a genetic screening blood test some kind of "invasion of privacy"? Some slight doubts appear (see Table 42). While more than one-half note that invasion of privacy definitely is not involved, there are some 40% who are more equivocal, observing either that it is *probably not* an invasion of privacy, while 12% reflect that it

TABLE 40 Carrier's View Re Necessity of Screening: Is Genetic Screening Program, as for TSD (CA-33)

absolutely essential	(61)	82.4
desirable/not essential	(11)	14.9
not desirable/but can be tolerated	(1)	1.4
should be stopped	(0)	0
no ansewer	(1)	1.4
	(74)	100

TABLE 41 Carrier's View Re Whether Screening Should Be Compulsory?" Should Blood Test (genetic screening) Programs, Such as TSD Program Bė (CA-35)

compulsory	(27)	36.5
voluntary	(45)	60.8
don't know	(2)	2.7
	(74)	100

probably is an invasion of privacy in some way. This concern with privacy, representing a not inconsiderable subgroup within the carrier sample, would suggest a further emphasis on voluntary, not compulsory, genetic screening approach.

In the Jewish population generally, may we again expect a mostly liberal attitude toward abortion? As is shown in Table 43, abortion on request is favored by overwhelming majorities, among both carriers and their spouses. Endorsement is somewhat higher (82%) for carriers than for spouses (72%). Even those not favoring abortion on request are quite prepared to endorse abortion in specific cases, including a broad range of conditions relating to the mother's physical or mental health. Virtually no total objection to abortion is found (see Table 43). This pro-abortion attitude is generally consistent with findings in evidence for general Jewish population samples, and with findings as reported by Margolin

TABLE 42 Carrier's View Re Privacy: Is Genetic Screening Blood Test Invasion of Privacy (CA-42)

definitely is	(invasion of privacy)	(0)	0
probably is	(invasion of privacy)	(9)	12.2
probably is not	(invasion of privacy)	(21)	28.4
definitely is not	(invasion of privacy)	(41)	55.4
no answer		(3)	4.1
		(74)	100

TABLE 43 Attitude Toward Abortion (CA-23)

	Carrier (CA-23)		*Spouse (SP-15)*	
favor abortion/on request by the pregnant women	(61)	82.4	(53)	71.6
favor abortion/in specific cases	(12)	16.2	(19)	25.7
against abortion under any circumstances	(0)	0	(1)	1.4
no answer	(1)	1.4	(1)	1.4
	(74)	100	(74)	100

(1978) in a cross-sectional (not exclusively Jewish) university student population.

Considering the profile of the results in this Tay-Sachs disease carrier (and carrier spouse) follow-up study, it seems clear that the reported feelings and reactions do not lend credence to a view that diagnosis as a TSD carrier is perceived as stigmatizing. Rather, some sense of being troubled or concerned, the injection of a dynamic of anxiety affecting both carrier and spouse, appears in consequence. This turbulence, mostly moderate or mild—in perhaps one case in five severe (perhaps melded with other pre-existing family dynamics)—may take its subtle toll, but mainly in the face of denials, to the effect that carrier status has had little impact on plans to have children, on birth control, on sexual activity, and so on.

Whatever else, carriers and their spouses emerge as strong supporters of continuing genetic screening, particularly on a voluntary basis. And there continues the characteristic support for abortion, on general principles or as specific therapeutic intervention, consistent with other findings in Jewish populations in comparable inquiries. In this context, TSD carriers and their spouses, far from responding with fright or withdrawal, assert the validity of genetic screening efforts and show themselves ready to make use of drastic interventions, as may be warranted by circumstances.

As the initial impact of being a carrier is absorbed as time passes, the concern (or anxiety) level further declines. While 21.6% of carriers reported themselves retrospectively as deeply troubled when they first learned of their carrier status, there are none reporting feeling this way after about one year following the event. If one combines the deeply troubled and quite concerned categories, the total figure declines from 51.3% initially to 15.1% at the later time. (Still, on a relative basis, concern fades less for those 29 and under, especially if they are female, than for other carrier categories—again an indication of the salient function of life status, and the associated significance of psychological proximity to conception.)

INTERPRETATION

Returning to the initial schema, one may surmise that under present conditions of communication flow, life status, and certain core motivators, the identification of an individual as a TSD carrier does not act as a deterrent to potential use of newly developed genetic disease prevention technologies. In the framework of the interacting variables, any possible concern with stigma is amply cushioned or overridden by more salient forces that place the individual in a state of support (and, if necessary, possible direct use) of genetic screening and therapeutic abortion. Particularly noting the impact of education and reasonable genetics knowledge (shown to be high but not *exceptionally* high for the carrier sample, in data not reported here), and considering value systems espousing knowledge for its own sake as well as practical guides, the path remains clear for the carrier subpopulation to continue direct connection with and support for the genre of technologies considered. Surely, while identification as carrier may create some discomfort or lack of ease for the individual and in the family setting, there is no indication of attitude or behavior warranting characterization as stigma.

NOTE

1. In a follow-up of research of this nature, recall data play a particularly important and useful role. For a critical consideration of this methodology, see Spanier (1976).

Chapter 8

RATIONALE AND METHOD FOR A PROTOTYPIC STUDY

New developments in human genetic technology, rooted in advances in fundamental and applied genetic, cytologic, and biochemical research have become important factors in the prospective control of genetic disease. For instance, a significant advance in genetic technology is constituted by the present technical capability to identify more than sixty recessively inherited conditions, as well as nearly all cytogenetic abnormalities, in the second trimester fetus through analysis of amniotic fluid or amniotic fluid cells obtained by amniocentesis in early pregnancy. This procedure has been applied mostly in pregnancies at risk for chromosomal abnormalities in the fetus, and, in about 10% of instances, to pregnancies at risk for metabolic, genetically predictable inborn errors. Most recently these procedures have been applied to the evaluation of fetal neural tube abnormalities through the measurement of amniotic fluid-fetoprotein concentrations. A direct consequence of application of this genetic technology is the availability of accurate fetal diagnosis, and selective abortion if the fetus is found to be affected. By this means, many families have been given the opportunity to have healthy children who, otherwise, might never have been conceived.

As one assesses present and long-term applicability and impact of genetic technologies of this nature, one needs to become aware of the prevailing patterns of public understanding and misconception of genetic principles and human genetic disease. This is the case especially if large-scale programmatic applications are required, and if these applications are voluntary rather than legislatively mandated.

Further, these new technologies confront society with a plethora of unanswered ethical and legal questions. Such answers necessarily relate to the underlying values and attitudes held by people in communities and by the subpopulation actually or potentially affected by these emergent genetic procedures. Ferment in values and attitudes in the sixties and seventies may presage heightened readiness for change and for the wider utilization of the technologies, as discussed in Chapter 2. If it is assumed that such utilization depends to some extent on an adequate level of public understanding of the relevant issues, it becomes important to examine, as one of a number of interactive variables, the prevailing knowledge levels regarding genetic principles and human genetic disease in the subpopulations concerned. If insufficient public knowledge is found to exist, or if there is a purposeful or inarticulate resistance to utilization of the new developments, the latter are likely to lie fallow, or may come to be translated into programmatic form prematurely, with the result that considerable professional energies and allocated funds will be poorly deployed.

By way of example, one may wish to determine the extent to which an identifiable subpopulation will participate—on the basis of public information, persuasion, and preexisting personal motivation—in a communitywide genetic disease carrier screening program. The results of the screening provide important sets of benefits: indication of genetic disease carrier status and its possible reproductive consequences, opportunities for genetic counseling, review of options in mating choice, recommendation of amniocentesis where indicated, and ultimately consideration of selective abortion if the birth of a seriously abnormal or moribund child is forecast.

The prospective prevention of a rare recessive disorder, Tay-Sachs disease, represents perhaps the earliest model of an approach as described. Through heterozygote (carrier) screening in the population predominantly at risk for this condition—mainly Jewish persons of European ancestry—couples who may give birth to affected offspring can be identified. Such identification occurs if both parents are found to be carriers. Once identified, the availability of accurate and safe early prenatal diagnosis enables these couples to have only unaffected offspring, assuming that the parents choose to monitor each pregnancy by amniocentesis and selectively terminate pregnancies in which the fetus is found to be affected with Tay-Sachs disease.

Against this background, *prior* to the inception of a community-based education program and TSD carrier screening effort in Los Angeles, in 1973-1974, it was elected to assess genetic disease knowledge levels and relevant attitude profiles of the specified "target" subpopulation: childbearing-age Jewish individuals.

METHOD AND SAMPLE

A mail questionnaire survey procedure, with systematic follow-up of nonrespondents, was chosen to assess the levels of knowledge of genetic principles and of human genetic disease, and attitudes relating to genetic disease and its control in the target subpopulation. The effectiveness of this approach as a means for gauging the subpopulation's knowledge level (and other relevant psychosocial aspects) depends (a) on the appropriate delineation of and sample development in the target subpopulation, (b) proper specification of the key issues and design of appropriate questionnaire content and format, and (c) assurance that those responding

TABLE 44 Sample and Response

Source	*Question-naires mailed*	*Response recieved*	*Response rate (%)*	*Distribution of response by source (%)*
Jewish Federation list	1125	597	53.1	58.6
Distinctive Jewish names phonebook list	988	357	36.1	35.1
In response: source notation lacking	x	64	x	6.3
	2113	1018	48.2	100.0

reflect the relevant subpopulation's critical characteristics or minimally that the extent and direction of possible differences between respondents and nonrespondents be adequately ascertained. The survey methodology is briefly reported below.

The fourteen-page, one hundred and five-question survey questionnaire was developed on consultation with practitioners and scholars in human genetics and in sociopsychological and survey research. Differences between mail questionnaire respondents and nonrespondents were examined by means of an intensive telephone follow-up survey, focusing on a subsample of nonrespondents. The mail questionnaire survey phase was conducted between July and November 1973, preceding by six months the introduction of the formal Tay-Sachs disease community education program.

Two sources were used as the basis for sample selection: (a) household addresses provided by the Jewish Federation-Council of Greater Los Angeles (Jewish Federation), to include high proportions of couples in the childbearing ages, and (b) household addresses chosen from specified Greater Los Angeles urban and suburban telephone directories on the basis of a controlled list of "distinctive Jewish names" (DJN), widely used as an indicator of Jewish population distribution.

A total of 2113 individual questionnaires, with postage-paid return envelopes, were mailed, and a duplicate was sent after six weeks if no response was received. Responses arriving more than twelve weeks after the initial mailing were excluded, yielding 1018 usable responses–a response rate of 48.2% (see Table 44). A subset, n = 847, primarily eliminating older, nonreproductive age persons, is included in most analyses reported.

As anticipated, a higher response rate was gained from the Jewish Federation list (53.1%) as compared with that found for the distinctive Jewish names (DNJ) phonebook list, (36.1%).

Comparisons of respondents drawn from the two data sources indicate that Jewish Federation list respondents tend to be somewhat younger and more central to the mainstream of Jewish religious ideology than DJN phonebook respondents; (within childbearing age range 20-49: 39 and under Jewish Federal list 87.6% versus DNJ list 68.0% [X^2 sig .01]; college graduate or advanced degree: Jewish Federation list 65.4% versus DJN list 45.1% [X^2 sig .01]. Other comparisons showed no significant differences between the groups.

As to nonrespondents to the mail questionnaire: In a randomly chosen subsample, 130 persons who did not return either of the two mailed questionnaires were contacted by telephone, with as many as nine call-backs if required, and a short-form genetics knowledge question list was used to collect comparable data to the mail questionnaire. This follow-up demonstrated a significant tendency (X^2 sig .01) for nonrespondents to be older and less educated. *Thus one notes the higher tendency of persons in the younger age category to reply in the mail questionnaire study phase, a tendency consistent with the study's focus on a reproductive-age target population.* However, this difference between mail questionnaire respondents and nonrespondents needs to be kept in mind in interpretation of study findings.

RESPONDENT CHARACTERISTICS

As indicated, since the major intent is to characterize the genetics knowledge profile of individuals of childbearing age,

TABLE 45 Age and Sex

		% of sample	(n)
Males			
	20-29	9.3	(79)
	30-39	27.4	(232)
	40-49	11.3	(96)
Subtotal		48.0	(407)
Females			
	20-29	18.4	(156)
	30-39	27.4	(232)
	40-49	6.1	(52)
Subtotal		51.9	(440)
Total		100.0%	(847)

TABLE 46 Marital Status

	% of sample	(n)
Married	94.8	(803)
Single (never married)	3.1	(26)
Divorced, Widowed	2.1	(18)
Total	100.0%	(847)

NOTE; 171 of 1018 cases including sufficient data to warrant analysis, or 16.8% of the usable sample, are eliminated from the analysis because they fall outside the reproductive-age range, 20-49, as specified.

data reported are confined principally to those for persons ages 20-49 (n = 847; 83.2% of total usable responses received). Tables 45 and 46 summarize the salient characteristics of the study sample:

(1) The respondent sample is fairly evenly divided between males (48%) and females (52%).

TABLE 47 Jewish Religious Ideology (self-description)

	Males	Females	Total	National JPS, Household Heads, 1970
Orthodox	(10) 2.5	(8) 1.8	(18) 2.1	11.4
Conservative	(156) 38.3	(174) 39.5	(330) 39.0	40.4
Reform	(119) 29.2	(166) 37.7	(285) 33.7	30.0
Nonpracticing	(99) 24.3	(57) 13.0	(156) 18.4	15.6
Other	(23) 5.6	(35) 8.0	(58) 6.8	2.6
Total	(407) 100.0%	(440) 100.0%	(847) 100.0%	100.0%

X^2 (Males/Females) = 21.49 sig. .001; X^2 (Total/NJPS) = 115.43 sig. .001

SOURCE: Massarik, Fred, *Jewish Identity,* New York: Council of Jewish Federations and Welfare Funds, 1974, p. 2.

(2) About 55% in the respondent sample fall in the 30-39 age bracket.

(3) Proportionately more younger females (20-29), 18.4% of the sample, are included, compared to 9.3% males of corresponding age.

(4) Nearly 95% in the sample are married.

(5) Respondents have had extensive childbearing and childraising experience. The average (arithmetic mean) number of living children reported is 1.91. As specified by husband's age, the number of living children reported is 2.46 for the oldest age category, 40-49, and, as expected, lowest, .81 children, for the youngest category, 20-29.

(6) The sample respondents are predominantly identified (regardless of formal congregation membership) with the "mainstream" ideological positions of the American Jewish community. Those describing themselves as Conservative constitute the largest single segment, almost 40%, followed by Reform, about one-third. A somewhat higher proportion of females than males identifies as Reform, while the opposite holds for the nonpracticing. Men are more inclined to describe themselves simply as "just Jewish" without specific ideological preference. As con-

TABLE 48 Education Level (highest level attained)

	Males 20-29	30-39	40-49	Total	*U.S. National Jewish Population 1970*
High school / less	(1) 1.3	(10)[a] 4.3	(9) 9.4	(20) 4.9	39.0
Some college	(16) 20.3	(25) 10.8	(24) 25.0	(65) 16.0	18.0
College graduate	(30) 38.0	(78) 33.6	(34) 35.4	(142) 34.9	15.4
Professional degree[c]	(32) 40.5	(119) 51.3	(29) 30.2	(180) 44.2	27.6
	(79) 100.0%	(232) 100.0%	(96) 100.0%	(407) 100.0%	100.0%

X^2 (by age, rows 1&2 combined) = 19.58, df = 4, sig. .001
X^2 (Total/NJPS) = 257.94, df = 3, sig. .001

	Females 20-29	30-39	40-49	Total	
High school/ less	(9)[a] 6.6	(21)[b] 9.1	(9) 17.3	(39) 8.9	53.1
Some College	(51) 32.7	(98) 42.2	(22) 42.3	(171) 38.9	21.7
College graduate	(72) 46.2	(84) 36.2	(16) 30.8	(172) 39.1	14.2
Professional degree[c]	(24) 15.4	(29) 12.5	(5) 9.6	(58) 13.2	11.0
	(156) 100.0%	(232) 100.0%	(52) 100.0%	(440) 100.0%	100.0%

X^2 (by age, rows 1&2 combined) = 9.53, df = 4, sig. .05;
X^2 (Total/NJPS) = 403.36, df = 3 sig. .001
a. includes 1 case of less than high school graduation
b. includes 2 cases of less than high school graduation
c. any advanced degree beyond the Baccalaureate
d. individuals, age 25 up; A. Chenkin, National Jewish Population Study, *Demographic Highlights*, New York: CJFWF, 1973, p.9.

trasted with National Jewish Population Study data, the Orthodox appear less frequently in the respondent sample than among all Jewish household heads in the United States as a whole, probably a reflection of the younger age level of the respondent target group (see Table 47).

(7) The education level of respondents is exceptionally high, particularly for males. Almost 80% of men are college graduates or have attained an advanced professional degree; the correspond-

ing percentage for females is almost 50%. These percentages exceed corresponding National Jewish Population Study figures, in part again mirroring conditions prevalent in younger Jewish populations for whom secular education represents an important life goal and personal achievement (see Table 48).

(8) Respondents include a diversified range of occupations, with a variety of professions heavily represented (data not reported in statistical form).

The profile of the respondent sample well reflects the salient characteristics of the chosen *target* population: persons in the childbearing ages, married, and ethnically (though not always religiously) Jewish. By means of an economical mail questionnaire/plus telephone follow-up procedure, it becomes possible to obtain significant response from a geographically dispersed target subpopulation, as the basis for planning of a demographically focused genetic disease control program.

Chapter 9

INTERPERSONAL VALUES AND THE ETHICS OF GENETIC DISEASE CONTROL

Research findings as such do not, of course, provide answers to underlying questions of values and ethics. Indeed, the nature of the research process itself—the topics that are addressed, the paradigms chosen, the designs employed—these, as well as the utilization of knowledge, are reflective of implicit value positions and ethical orientations. Genetics disease control and sociopsychological research in this field themselves are both manifestations and crucibles of ethical concern.

The issue before us cuts across conventional fields of knowledge: for instance, the *Humanities Report,* published by American Association for the Advancement of the Humanities noted in early 1980 that

> current concern over the role of the humanities is tied closely to the rapid expansion, during the past two decades, of scientific knowledge and technology which has produced dramatic advances in both diagnosis and therapy. The technology explosion has been attended by a parallel explosion in bioethical questions.

Further evidence of this burgeoning interest is shown in the following: The Kennedy Institute at Georgetown University in recent years developed a Center for Bioethics, and the *Encyclopedia of Bioethics* has been published in four comprehensive volumes containing over 300 articles, covering an impressive range of topics, from "abortion" to "zygote banking." The University of Chicago Press publishes *The Journal of Medicine and Philosophy,* with frequent attention to the relationship between personal and interpersonal values and medical science and practice. In Spring 1978, the *National Forum,* the journal of honor society Phi Kappa Phi, devoted a complete issue to "Ethical Issues in Modern Medicine." Similar cross-disciplinary concerns are legion.

In a broad context, but beyond this monograph's purview, are recurrent concerns, within and outside academia, relating to "gene-splicing" and "recombinant DNA." Whether or not substantially founded in scientific fact, public unease and anxiety prevail regarding creation of new and perhaps dangerous forms of life, not to mention fears of more fantastic metamorphoses of "monsters" and hypothetical "Frankensteins." Considerable effort has been expended—as, for instance, in Cambridge, Massachusetts in the late 1970s and elsewhere—in public hearings and development of specific experimental guidelines for research in the field (see, for instance, Science, 1977).

Widespread focus on the ethics of research in recombinant DNA, a topic of substantial complexity and intrinsic significance, establishes a backdrop to ethical considerations associated with genetic disease control. Similar comments may be made regarding in vitro fertilization and the practical realization of the "test-tube baby."

Let us focus on the pivotal building-block of adult genetic screening, as illustrated by Tay-Sachs disease prevention programs. Given the empirical finding, consonant with widely prevailing social values, preferring *individual* choice with regard to participation or nonparticipation in genetic screening (with much lesser readiness to assign this responsibility to the medical profession, to the community, or to state government), it is the person's *voluntary act* that becomes central in this context. An act of this nature mirrors increasing personal and social responsiveness to humanistic values, to the person's intentionality and capacity for assuming responsibility for his/her action or inaction, and the resulting consequences. This orientation necessarily rests on some notion of *informed consent.* Such consent derives meaning from two sources (a) the person's inherent *capability* to bestow or to withhold such consent, and (b) the process of being or of becoming informed by means of learning, through various approaches to such learning, such as formal education in school and university, mass media, and special information campaigns focused specifically on genetic disease control.

The notion that the individual human being is and *ought to be* able to determine his/her destiny, i.e. to consent to one or another course of action, by choice and with a purposeful orientation, establishes a basis for a system of ethics. With individual value affirmed, the *person* becomes "the measure of good," rather than, for instance, the will of the state, or the welfare of the community sui generis. The concept of being informed asserts the beneficient value of knowledge, and implies a causal link between knowledge and the individual's personal and interpersonal good. These positions are, of course, relativistic—though not exclusive to the cultural milieus of the West and of a modern Westernized East. At other times, and in other places in the contemporary world as well, quite opposite value positions are held: the good of the state or community, or of specified religious precept are regarded as paramount, and as the summum bonum.

Further, these contrary views hold that much knowledge of certain kinds may well be dangerous and that a *little* knowledge as well may be a dangerous thing. Accordingly, it may be asserted that many important decisions relating to human welfare (whether individual in its own right, or for the larger systems of state, community, or religion) is best determined by the voice—and choice—of authority. Even within the modern Western setting—and, indeed, with some sense of realism—the role of "firm father," or at least that of expert is upheld, as in the trust of the individual in the physician's judgment.

The idea of informed consent, as may underlie voluntary participation in genetic screening, is fraught with complex and often unrecognized dilemmas. The matter goes beyond conformity to "human subjects" guidelines and beyond the signing of consent forms, prior to implementation of a medical intervention. One must ask: "To what extent does the person consenting truly *understand* the implications of her/his action?" And "Under what conditions of freedom and constraint is consent obtained?"

Clearly, understanding may range from superficial knowledge of a few "facts" to a well-internalized grasp of the subtleties of a procedure. The very signing of a document purporting to obtain informed consent may be viewed as nothing more than another routine bureaucratic procedure, or as a piece of paper whose small print either is irrelevant or constitutes a minimum condition for obtaining some desired response. One readily imagines an authoritarian regime, or even a tightly run benign bureaucracy, that by the creation of certain social and environmental constraints creates an apparent plethora of "the informed consenting."

In practical terms, the implementation of informed consent in genetic disease screening and in other approaches to genetic disease control, requires minimally (a) a clarification of what is *meant* by informed consent, considering just what it is that the person participating is to know, and how she/he perceives the substance of providing the consent noted; (b) how this consent is to be obtained, attending to the *system of freedom-and-constraint* within which the process of consenting moves forward—noting possible operation of subtle pressure or implicit coercion, in contrast with broad individual consideration of options and their alternate consequences, and (c) consideration by the consenting person's of the booby traps that might emerge as consequences even in well-intentioned genetic disease control procedures—as, for instance, potential generation of anxiety, psychological limitations on choice of marital partner, and so on. This study's results, however, suggest that with *meaningful* consent given by persons in a highly educated and generally knowledgeable population, with care in avoidance of undesirable implicit or explicit coercion, "anxiety costs" are relatively modest and transitory.

Another issue of ethic relevance is the *right of privacy.* Unless the person participating in the genetic disease control effort autonomously *wishes* to share information regarding his/her participation with others, one may assert that the participant's privacy, as relates to choice of participation and results, must be protected. Again, this view is embedded in a humanistic value position, assigning to the *person* the option of disclosure or nondisclosure of potentially sensitive individual actions.

Such right of privacy—a topic of considerable public concern in recent years, particularly with the advent of computer "data banks" and other large-scale methods for information storage and retrieval—is not confined exclusively to the revelation of test *results,* as may be derived in the course of blood test screening, or for that matter by intrauterine diagnosis. The simple physical act of participation, being present at a screening *with others,* entails a certain readiness by the participant to yield privacy—that is, to *publicly* affirm her/his interest in possible carrier status and related issues. The program sponsor's delimitation of concern with possible stigma, by positive education programs and by other means to assert the responsible and hygienic character of such participation, represents efforts to facilitate this specific public affirmation and its correlative compromise of absolute privacy. Indeed, readiness to "go public," for instance as regards ethnic identity, may be a significant aspect of participation in certain kinds of genetic screening programs, as a modicum of affirmation of Jewishness in Tay-Sachs disease screening. In this sense, a fine balance exists between the person's privacy on one hand and his/her readiness to affirm in a public setting, such as a TSD screening at a visible location, concern with genetics and ethnic identity.

In spite of the widespread success of TSD screening programs by community involvement, indicated by the number of persons participating, often in open physical settings such as synagogues, community centers, and universities, questions have been raised regarding ethics and values in this procedure. For instance, Motulsky (1974; italics added) asserts that, in TSD screening, "The approach used in these programs was to get all members of the community *aroused* about a disease which affects one/4000 births." In proposing "more medically oriented schemes," in contrast to those with a community or public orientation, Motulsky argues, without using the term, for a more specifically *non*public position.

While it may be a simple and relatively noncontroversial matter to assure privacy in the release of test results—a correlary of individual voluntary action—upon further consideration the matter quickly becomes more complex. For instance, if in genetic disease screening the finding that a given person is a carrier has implications for the possible reproductive destinies of that person's relatives, does the physician—or anyone else for that matter—have a right, or perhaps even the responsibility, to actively strive to inform these relatives of the circumstances? What if the choice is left to the individual and he/she, exercising voluntary action and right of privacy, chooses *not* to reveal this information to anyone, including those who might be directly or indirectly affected? Should some external authority—whether physician, health department, or other governmental agency—step in? And what are the tradeoffs? The dimensions of the ethical dilemma soon are manifest.

Then there is the matter of *wrongful life.* In essence, this concept, as recently subject to diverse legal rulings, holds that, under specified conditions the very fact of having been born may result in such pain and suffering both to the person born himself/herself and to others such as parents, suit may be brought in a court of law to secure amelioration and compensation for such pain and suffering and other grievances. The issue is complex. Particular defendants may be held responsible for the creation of this wrongful life, as in a child born with TSD. Such defendants may include laboratories who may have allegedly erred in preparation of test results, physicians who may have failed to secure relevant genealogical data, and others who purportedly may be at fault in this connection.

Whether or not it is possible to compare—at least in legal terms—*any* condition of life with a condition of *non*life is an issue in flux. In at least one instance (California State Court of Appeals, June 1980) it has been ruled that a suit of this nature can be brought, though in prior instances, rulings on this issue had been negative. At this reading, circumstances may have changed further, by diverse judgments and by varied interpretations. At the heart of the matter stands that ancient conundrum of meaning and value of life, in this instance contrasted with the assertedly preferred condition of a given (genetically afflicted) individual hypothetically remaining unborn. It becomes a matter of considering whether life-with-defect is in any sense in the same realm of discourse as nonlife, and in the range of torts, whether damages may be sought for economic burden created, funeral costs, and pain and suffering of the survivors. As more genetic testing is implemented, covering a greater variety of diseases with the afflicted fetus potentially therapeutically abortable, the issue of wrongful life will become of increasing ethical and legal import. Kaback and Zieger (n.d.), Motulsky (1974), and Moraczewski (in Sperber and Jarvik, 1976: 163-182) have discussed, from a variety of perspectives, the ethical issues relating to *genetic counseling.* Essentially genetic counseling is the process in which the implications of a particular genetic test result or similar assessment are purposefully explored with the person(s) involved. While in some instances such counseling has been carried out directly by the physician, in others trained genetic counselors are assigned to this function.

Information itself—including information on genetic test results—is not neutral nor is the process in which such information is communicated. And the manner in which information is interpreted clearly flows from a complex interpersonal process, reflecting preexisting personality and social conditions in the counselee and the personality and style of the counselor. The simple homilies help little: To speak of the "doctor-patient relationship" in genetic counseling, even if the counselor is a physician, or to allude to "nondirective counseling methods" serves only as the roughest guideline. While one may emphasize the need for confidentiality and for autonomy in decision-making by the counselee, subtle processes operate. For instance, though the counselor may be intellectually committed to the counselee's right to determine what to do by unaware cues, *how* things are said, facial expressions and other nonverbal clues, he/she may unconsciously reflect his/her own preference for therapeutic abortion as *the* solution. Alternatively, the genetic counselor may respond to the counselee's urgent request for specific recommendation, though initially he/she may have been committed to the counselee's autonomous decision-making process.

Opting for the nondirective stance once more underlines belief in the value of individuality. Yet acquiescing to the counselee's apparent need for structure, for specific advice, may be humanistically responsive to meeting that person's or family's psychological needs for guidance. An often implicit tug of war among conflicting value ascriptions comes into play.

Finally, the perennial and literally vital issue of abortion requires mention. The research results reported make clear the study respondents' prevalent readiness to choose therapeutic abortion, especially when the fetus is afflicted with genetic disease. The results demonstrate, at least at the level of hypothesis, that even those who are committed to ostensibly orthodox religious positions currently are potentially accepting of abortion as a means for prevention of birth of an afflicted child. These findings suggest that *ethics-in-practice* are subject to shifts with changing times and circumstance. Learned discussion of the nature of human life and of the presumed beginnings of the human person qua person continue. These consider, for example, the extent to which the fetus is active (human?), or the extent to which the fetus is regarded as representing potentiality and thus potential personess. (See, for instance varied approaches in Group for the Advancement of Psychiatry (1970) and Wade (1975).

Though the term "situation ethics" is no longer as popular as it was in the mid-1970s, the fact remains that the realities of ethics are defined by what people think and do under specified conditions. In more esoteric perspective, one may wish to argue for absolutes. Indeed, such firm assertions, themselves changeable in the long run, provide important bench marks in the understanding of real-life manifestations of ethical and nonethical behavior. Still, if we are to address purposefully the challenges and dilemmas created by the new technologies that have become available in the control of genetic disease, we must increasingly direct our attention to the values held by the human beings who are to benefit by the effective application of these significant advances.

Appendix A

QUESTIONNAIRE ITEMS/MAIL SURVEY/CALIFORNIA TAY-SACHS DISEASE PREVENTION PROGRAM/A PROTOTYPE OF PSYCHOSOCIAL GENETICS RESEARCH

SECTION I–TAY-SACHS DISEASE

1. HAVE YOU EVER HEARD OF TAY-SACHS DISEASE?
   Never_____ Something ______ A lot ______

2. HAVE YOU EVER KNOWN A FAMILY WHERE TAY-SACHS DISEASE OCCURRED?
   Yes ______ No ______ Don't know ______

3. HAS TAY-SACHS DISEASE EVER OCCURRED IN YOUR FAMILY?

Yes _____ No _____ Don't know _____

4. IF YOU HAVE HEARD OF TAY-SACHS DISEASE, WHERE DID YOU OBTAIN YOUR INFORMATION?

Clergyman	Yes _____	No _____
Friends	Yes _____	No _____
Magazine	Yes _____	No _____
Newspaper	Yes _____	No _____
Physician	Yes _____	No _____
Radio	Yes _____	No _____
Relatives	Yes _____	No _____
School	Yes _____	No _____
Scientific Literature	Yes _____	No _____
T.V.	Yes _____	No _____
Other (please specify)	Yes _____	No _____
_______________	Yes _____	No _____

5. TO THE BEST OF YOUR ABILITY RESPOND TO EACH OF THE FOLLOWING STATEMENTS (do not consult any source of information):

A. Symptoms of Tay-Sachs disease are evident at birth

Yes _____ No _____ Don't know _____

B. Tay-Sachs disease only occurs in Jewish children

Yes _____ No _____ Don't know _____

C. Any couple (regardless of race, religion or origin) can have a child with Tay-Sachs disease

Yes _____ No _____ Don't know _____

D. Tay-Sachs disease is a rare disease even in Jews

Yes _____ No _____ Don't know _____

E. Tay-Sachs disease is always fatal

Yes _____ No _____ Don't know _____

F. For a couple's children to be at-risk for Tay-Sachs disease, only *one* parent must be a carrier of the Tay-Sachs gene

Yes _____ No _____ Don't know _____

G. For a couple's children to be at-risk for Tay-Sachs disease, *both* parents must be carriers of the Tay-Sachs gene
Yes ______ No ______ Don't know ______

6. HAVE YOU EVER HEARD ABOUT A SCREENING PROGRAM FOR TAY-SACHS DISEASE PREVENTION?
Yes ______ No ______
A. If you answered *YES* to question 6, where did you hear about it?

In Los Angeles	Yes ______	No ______
Other city	Yes ______	No ______

SECTION II–GENETICS IN GENERAL

1. WHILE ILLNESS CAN HAVE MANY CAUSES (INFECTION, TRAUMA, MALIGNANCY), CERTAIN DISEASES ARE DUE *ALMOST ENTIRELY* TO INHERITED FACTORS. WHAT PERCENTAGE OF ALL ILLNESSES DO YOU THINK ARE PREDOMINANTLY DUE TO INHERITED FACTORS (GENETIC CAUSES)? (select only one answer)
Less than 1% ______
5% ______
10% ______
25% ______
50% ______
70% or more ______

2. IS AN "INHERITED DISORDER" THE SAME AS A "CONGENTIAL DISORDER"?
Yes ______ No ______ Don't know ______

3. A "GENE" IS BEST DEFINED AS (check one):
The chromosome ______
A chemical unit of hereditary information ______
A group of cells with hereditary information ______
The egg and the sperm ______
Don't know ______

4. RESPOND TO EACH OF THE FOLLOWING STATEMENTS CONCERNING GENETIC DISEASES:
   A. Certain genetic diseases can skip generations
      Yes ______ No ______ Don't know ______
   B. Certain genetic diseases can be inherited from *one* parent only
      Yes ______ No ______ Don't know ______
   C. Certain genetic diseases require inheritance of the disease gene from *both* parents
      Yes ______ No ______ Don't know ______
   D. Most cases of genetic disease are due to a new mutuation in the affected individual
      Yes ______ No ______ Don't know ______

5. DO YOU BELIEVE THAT GENETIC DISEASES CAN BE CURED?
   Always_____ Sometimes_____ Never_____ Don't know_____

6. DO YOU BELIEVE THAT GENETIC DISEASES CAN BE TREATED?
   Always_____ Sometimes_____ Never_____ Don't know_____

7. DO YOU BELIEVE THAT GENETIC DISEASES CAN BE PREVENTED?
   Always_____ Sometimes_____ Never_____ Don't know_____

8. DO YOU BELIEVE THAT GENETIC DISEASES CAN BE PREDICTED?
   Always_____ Sometimes_____ Never_____ Don't know_____

9. CAN A PERSON CARRY A DISEASE-RELATED GENE AND STILL BE PERFECTLY HEALTHY?
   Always_____ Sometimes_____ Never_____ Don't know_____

10. CAN A GENETIC DISORDER OCCUR ALMOST EXCLUSIVELY IN MALES?
    Yes ______ No ______ Don't know ______

11. CAN A GENETIC DISORDER OCCUR ALMOST EXCLUSIVELY IN FEMALES?
    Yes ______ No ______ Don't know ______

12. CAN A GENETIC DISORDER, OTHER THAN TAY-SACHS DISEASE, OCCUR ALMOST EXCLUSIVELY IN A SELECTED POPULATION (RACIAL, ETHNIC, OR NATIONAL GROUP)?
    Yes ______ No ______ Don't know ______

13. HOW OFTEN IS A GENETIC DISEASE LIKELY TO OCCUR IN A SELECTED POPULATION?
    Frequently ______ Occasionally ______Very rarely ______

14. IF A COUPLE IS TOLD THAT THEY HAVE A 25% RISK FOR HAVING CHILDREN WITH A GENETIC DISEASE THIS WOULD MEAN (answer each statement):
    A. that none of their children would be affected
       Yes ______ No ______ Don't know ______
    B. that each child would be 25% affected
       Yes ______ No ______ Don't know ______
    C. that if the first child was affected, the next three would be entirely unaffected
       Yes ______ No ______ Don't know ______
    D. that each pregnancy would have a one in four chance of resulting in an affected child
       Yes ______ No ______ Don't know ______

15. WHICH OF THE FOLLOWING CONDITIONS WOULD YOU CONSIDER ARE PREDOMINANTLY GENETIC DISEASES (STRONGLY DUE TO INHERITED FACTORS)?
    A. Allergy Yes____ No____ Don't know____
    B. Asthma Yes____ No____ Don't know____
    C. Cancer Yes____ No____ Don't know____
    D. Color blindness Yes____ No____ Don't know____
    E. Cystic fibrosis Yes____ No____ Don't know____
    F. Diabetes Yes____ No____ Don't know____
    G. Dwarfism Yes____ No____ Don't know____

H.	German measles	Yes____	No____	Don't know____
I.	Gonorrhea	Yes____	No____	Don't know____
J.	Heart disease	Yes____	No____	Don't know____
K.	Kidney disease	Yes____	No____	Don't know____
L.	Leukemia	Yes____	No____	Don't know____
M.	Mental illness	Yes____	No____	Don't know____
N.	Mongolism	Yes____	No____	Don't know____
O.	Near-sightedness	Yes____	No____	Don't know____
P.	Pneumonia	Yes____	No____	Don't know____
Q.	Rheumatic fever	Yes____	No____	Don't know____
R.	Sickle cell anemia	Yes____	No____	Don't know____
S.	Tay-Sachs disease	Yes____	No____	Don't know____

16. IF A SIMPLE BLOOD TEST COULD DETERMINE IF AN INDIVIDUAL HAS THE POTENTIAL FOR A GENETIC DISEASE IN HIS OR HER OFFSPRING, UNDER WHICH OF THE FOLLOWING CIRCUMSTANCES WOULD YOU BE WILLING TO TAKE SUCH A TEST?

A. If the disease could be effectively treated
Yes ______ No ______ Don't know ______

B. If nothing could be done for the disease (and therefore the test would be for information only)
Yes ______ No ______ Don't know ______

C. If the disease were untreatable, but could be detected early in pregnancy and abortion offered if the fetus were affected with the condition
Yes ______ No ______ Don't know ______

EARLY PREGNANCY WHEN PREGNANCY COULD STILL BE TERMINATED, WHO SHOULD PLAY THE MAJOR ROLE IN THE DECISION TO HAVE OR NOT TO HAVE THIS TEST? (respond to each of the following)

A. The state government, by requiring such a test by law
Yes ______ No ______ Don't know ______

B. Each community (religious or ethnic group) where this genetic disorder is most likely to occur
Yes ______ No ______ Don't know ______

C. The medical profession, by instructing patients to take such a test
Yes _____ No _____ Don't know _____

D. The individual couple, making the decision independent of outside influences
Yes _____ No _____ Don't know _____

SECTION III–GENERAL ATTITUDES

1. WHICH OF THE FOLLOWING STATEMENTS BEST REFLECTS YOUR ATTITUDE TOWARD ABORTION? (check one)
   A. I favor abortion if requested by the pregnant woman. _____
   B. I am against abortion under any circumstances _____
   C. I favor abortion in specific cases. _____

2. IF YOU CHECKED "C," UNDER WHICH OF THE FOLLOWING CIRCUMSTANCES WOULD YOU FAVOR ABORTION? (check as many as apply)
   A. If the mother's physical health is in danger _____
   B. If the mother's mental health is in danger _____
   C. If the size of their family already exceeds their economic capacity. _____
   D. If the mother is a teenager. _____
   E. If the mother is unmarried. _____
   F. If the mother is over the age of forty. _____
   G. Other (please specify)

3. WOULD YOU FAVOR ABORTION IF IT COULD BE DETERMINED EARLY IN PREGNANCY (3-4 MONTHS) THAT THE CHILD RESULTING FROM THAT PREGNANCY WOULD *DEFINITELY* HAVE:
   A. Slight physical handicap
   Yes _____ No _____ Don't know _____
   B. Severe physical handicap
   Yes _____ No _____ Don't know _____
   C. Mild mental retardation
   Yes _____ No _____ Don't know _____

D. Severe mental retardation
Yes ______ No ______ Don't know ______

E. A disease leading to death before age 5
Yes ______ No ______ Don't know ______

F. A disease leading to death before age 10
Yes ______ No ______ Don't know ______

G. A disease leading to death before age 20
Yes ______ No ______ Don't know ______

SECTION IV–RESPONSIBILITIES

Medical science is making dramatic progress in the detection and prevention of disease. Increased knowledge means increased responsibility for mankind. We would like to know your personal feelings about some of the questions which may arise with this new facet of health care. Therefore, please give special consideration to the following questions and answer as indicated.

1. IF PARENTS COULD TAKE A SIMPLE BLOOD TEST BEFORE PREGNANCY TO ESTABLISH THEIR RISK OF EVER HAVING A CHILD WITH A SERIOUS GENETIC DEFECT, WHO SHOULD PLAY THE MAJOR ROLE IN THE DECISION TO HAVE OR NOT TO HAVE THIS TEST? (respond to each of the following)

   A. The state government, by requiring such a test by law
   Yes ______ No ______ Don't know ______

   B. Each community (religious or ethnic group) where this genetic disorder is most likely to occur
   Yes ______ No ______ Don't know ______

   C. The medical profession, by instructing patients to take such a test
   Yes ______ No ______ Don't know ______

   D. The individual couple, making the decision independent of outside influences
   Yes ______ No ______ Don't know ______

2. SUPPOSE ANOTHER TEST WERE AVAILABLE WHICH WOULD ENABLE PHYSICIANS TO DIAGNOSE THE PRESENCE OF GENETIC DISEASE IN THE FETUS DURING

3. IF A SIMPLE BLOOD TEST WERE AVAILABLE TO DETERMINE A COUPLE'S RISK FOR A SERIOUS GENETIC DEFECT IN THEIR OFFSPRING (FOR WHICH THERE WOULD BE *NO* MEANS OF PREVENTION OR TREATMENT), DO YOU FEEL A COUPLE SHOULD TAKE SUCH A TEST?

   Yes ______ No ______ Don't know ______

   IF SO, WHEN?

   A. Before the time of engagement ______
   B. Before marriage ______
   C. After marriage, but prior to the planning of a family ______
   D. During the early part of pregnancy, when a second test (described in # 2, above) can determine if the fetus is affected or not ______
   E. Other (please specify)
      ______________________ ______

4. IF A SIMPLE BLOOD TEST WERE AVAILABLE TO DETERMINE A COUPLE'S RISK FOR A SERIOUS GENETIC DEFECT IN THEIR OFFSPRING (FOR WHICH PREVENTION IS POSSIBLE THROUGH FETAL DIAGNOSIS AND SELECTIVE ABORTION, THEREFORE ENABLING THAT COUPLE TO STILL HAVE HEALTHY CHILDREN), DO YOU FEEL A COUPLE SHOULD TAKE SUCH A TEST?

   Yes ______ No ______ Don't know ______

   IF SO, WHEN?

   A. Before the time of engagement ______
   B. Before marriage ______
   C. After marriage, but prior to the planning of a family ______
   D. During the early part of pregnancy, when a second test (described on the preceding page, # 2) can determine if the fetus is affected or not ______
   E. Other (please specify) ______

SECTION V–PERSONAL DATA

1. YOUR ZIP CODE: ______________________

2. SEX: Male ______ Female ______

3. AGE: ______

4. MARITAL STATUS:
   Single ______ Engaged ______
   Married ______ Divorced or Widowed ______

5. WHAT IS YOUR RELIGION (by birth)?
   Jewish____Protestant______Catholic______Other_____

6. WHAT IS YOUR CURRENT RELIGIOUS PREFERENCE?
   Jewish____Protestant______Catholic______Other_____

7. IF JEWISH, ARE YOU
   Orthodox ______ Conservative ______
   Reform ______ Non-practicing ______

8. WHAT IS YOUR SPOUSE'S RELIGION (by birth)?
   Jewish____Protestant______Catholic______Other_____

9. WHAT IS THE HIGHEST LEVEL OF EDUCATION YOU HAVE COMPLETED?
   Less than 12 years ______
   High school graduate ______
   Some college ______
   College graduate ______
   Professional degree ______

10. USUAL OCCUPATION (fill in): ______________________

11. FAMILY FACTS (for women only):

Number of pregnancies: ______
Number of living children: ______
Number of children (by age):
- Less than 5 years: ______
- 5-9 years: ______
- 10-18 years: ______
- Over 18 years: ______

12. FAMILY FACTS (for men only):

Number of living children: ______
Number of children (by age):
- Less than 5 years: ______
- 5-9 years: ______
- 10-18 years: ______
- Over 18 years: ______

Appendix B

PARTIAL QUESTION LIST TSD CARRIER

FOLLOW-UP INTERVIEW

Carrier

Thank you for giving us this opportunity to talk with you. I understand that sometime around ((MONTH/YEAR)) ____________ _____ you found out that you were a TSD carrier. I'd like to talk with you about that ((OPEN)). First, I'd like to ask a few questions about you and your family:

Q. 3. What is your religion by birth?

Jewish	_____
Protestant	_____
Catholic	_____
Other (which?) ____________	_____

Q. 4. Whether or not you now belong to a congregation, do you consider yourself

Orthodox	(Jewish)	_____
Conservative	(Jewish)	_____
Reform	(Jewish)	_____
"Just Jewish"	(without other preference)	_____
Other (which? incl. NJ) ________________________		_____

Q. 12. Thinking back once more to the day *when you first found out* that you were a TSD carrier . . . which of the following statements *then* best reflected your reactions?

A ______
B ______
C ______
D ______

Q. 13. Which of these statements best describes how your spouse felt *then* . . . when he/she first found out that you were a TSD carrier?

A ______
B ______
C ______
D ______

Q. 15. A bit more about what various people may think. Here is a card that shows some attitudes of people who might know about your being a TSD carrier. How did ((EACH SPECIFIED PERSON)) react? ((PROBE RE 'DOES NOT KNOW' AND 'MULTIPLE OPTIONS')): (several answers per person OK)

	does not know	con-cern, anxi-ety	approv-al re finding out	little feel-ing	shock	neu-tral	does not apply "no such"	other
A. Spouse	___	___	___	___	___	___	___	___
B. Child(ren)	___	___	___	___	___	___	___	___
C. Mother	___	___	___	___	___	___	___	___
D. Father	___	___	___	___	___	___	___	___
E. Brother(s)	___	___	___	___	___	___	___	___
F. Sister(s)	___	___	___	___	___	___	___	___
G. Other relatives? (which?)______	___	___	___	___	___	___	___	___
H. Close friend(s) of same sex	___	___	___	___	___	___	___	___
I. Close friend(s) of opposite sex	___	___	___	___	___	___	___	___

J. Work associate(s)	___	___	___	___	___	___	___	___
K. Regular physician	___	___	___	___	___	___	___	___
L. Other physician (e.g., specialist)	___	___	___	___	___	___	___	___
M. Psychotherapist	___	___	___	___	___	___	___	___
N. Professional counselor (other than psycho-therapist)	___	___	___	___	___	___	___	___
O. Somebody/anybody else who matters	___	___	___	___	___	___	___	___

Q. 19. Do you plan to have any (additional) children? (can be interviewer's recap of earlier answer):

yes ______
no ______
don't know ______
((IF 'YES')): How many more? ______

Q. 20. Are your plans about having/not having (additional) children in any way affected by your knowledge that you are a TSD carrier?

Not affected at all ______
Affected somewhat . . . it's one of several factors ______
Affected importantly . . . it is the major factor ______
Don't know ______
Other (which?) ____________________ ______

Q. 21. A. As far as your spouse is concerned, does (he)/(she) want to have any (additional) children?

yes ______
no ______
don't know ______

B. Would you say that you and your (husband)/(wife) *agree* or *disagree* on the plans for (not having) (having more) children? Or have you not discussed this?

agree ______
disagree ______
issue has not been discussed ______

C. ((IF DISAGREE)) What is your disagreement?

____

Q. 22. Do you believe that your spouse's attitude toward having (additional) children is affected by (his)/(her) knowledge that you are a TSD carrier?

Not affected at all ______
Affected somewhat . . . it's one of several factors ______
Affected importantly . . . is the major factor ______
Don't know ______
Other (which?) ______________________

Q. 23. Which of the following statements best reflects how you feel about abortion? Please choose *one* statement (A or B or C) only:

A. I favor abortion if requested by the pregnant woman ______
B. I am against abortion under any circumstances ______
C. I favor abortion in specific cases ______

Q. 33. Do you believe that a genetic screening program, such as the one for TSD

is absolutely essential ______
is desirable, but not essential ______
is not desirable, but can be tolerated ______
should be stopped altogether ______

((PROBE: WHY DO YOU FEEL THIS WAY?))

Q. 35. Do you believe that blood test screening programs (such as the TSD program) should be compulsory (that is everyone should be required to take the test), or that they should be *voluntary* (that is everyone should make up their own minds whether they want to take the test):

compulsory ______

voluntary ______

don't know ______

Q. 38. Do you believe that the publicity received by TSD

Is a kind of "stigma" on the Jewish people . . . it singles out Jewish people in some negative way ______

Is not a "stigma" on the Jewish people; it's ok to have such publicity ______

Never thought about it ______

Have no opinion/don't know ______

Q. 41. If someone–generally in ages 20-49–were to ask your advice concerning the desirability of their having a blood test at a TSD screening, what would you advise?

Definitely *stay away* from such a blood test . . . it's not a good idea ______

Consider having such a blood test . . . but with some doubts or concerns _____

((PROBE: WHAT DOUBTS?))

Probably have the blood test if it's convenient ______

Definitely have the blood test, under all circumstances, convenient or not ______

Would not want to advise (even if pressed) ______

Q. 42. Do you believe that asking a person to take a blood test for purposes of genetic screening is a kind of invasion of privacy?

Definitely is (invasion of privacy) ______
Probably is (invasion of privacy) ______
Probably is not (invasion of privacy) ______
Definitely is not (invasion of privacy) ______

Q. 47. Have the results of the TSD screening changed in any way, your feelings about birth control?

yes ______
no ______

Q. 48. Have the results of the TSD screening changed to any extent the frequency of your sexual activity?

increased frequency ______
no change ______
decreased frequency ______

Q. 54. Do you now seek medical attention more often or less often than before you found out that you are a TSD carrier?

more ______
no change ______
less ______
not now/not before ______

Spouse

Q. 6. Thinking back for a moment to the day your spouse *first found out* that he/she was a TSD carrier . . . which of the following statements *then* best reflected his/her reactions?

A ______
B ______
C ______
D ______

Q. 7. Which of these statements best describes how *you* felt *then* . . . when he/she first found out that you were a TSD carrier?

A ______
B ______
C ______
D ______

Q. 12. Do you plan to have any (additional) children?

yes ______
no ______
don't know ______

((IF 'YES')): How many more? ____________

Q. 14. Would you say that you and your (husband)/(wife) *agree* or *disagree* on the plans for (not having) (having more) children? Or have you not discussed this?

agree ______
disagree ______
issue has not been discussed ______

((IF DISAGREE)): What is the disagreement?

REFERENCES

BECK, E., S. BLAICHMAN, C. R. SCRIVER, and C. L. CLOW (1974) "Advocacy and compliance in genetic screening." New England Journal of Medicine (November 28).

BECKER, M. H. and L. A. MAIMAN (1975) "Social-behavioral determinants of compliance with health and medical care recommendations." Medical Care 13 (January).

BECKER, M. H., R. H. DRACHMAN, and J. P. KIRSCHE (1974) "A new approach to explaining sick-role behavior in low-income populations." American Journal of Public Health 64 (March).

BECKER, M. H., M. M. KABACK, I. ROSENSTOCK, and M. V. RUTH (1975) "Some influences on public participation in a genetic screening program." Journal of Community Health 1 (Fall).

CHENKIN, A. (1972) "Demographic highlights," in CJFWF, National Jewish Population Study. New York: Author.

CHILDS, B. (1972) Contribution, pp. 373-403 in Annual Review of Medicine, Volume 23.

--- L. GORDIS, M. M. KABACK, and H. H. KAZAZIAN, Jr. (1976) "Tay-Sachs screening: social and psychological impact." American Journal of Human Genetics: 550-558.

--- (n.d.) "Tay-Sachs screening: motive for participating and knowledge of genetics and probability." American Journal of Human Genetics 28 (November).

CLOW, C. L. et al. (1973) "On the application of knowledge to the patient with genetic disease." Progress in Medical Genetics 9.

Committee for the Study of Inborn Errors of Metabolism (1975) Genetic Screening: Programs, Principles, and Research. Washington, DC: National Academy of Sciences.

DAVIS, L. G. [ed.] (1975) Bibliographic Survey on Sickle Cell Anemia. Monticello, IL: Council of Planning Librarians.

DEVEY, G. B. and P.N.T. WELL (1978) "Ultrasound in medical diagnosis." Scientific American (May): 98-112.

FREITAS, R. A. (1980) "Fetal abortion: a technological solution to the problem of abortion ethics." Humanist (May/June).

FROST, N. and M. M. KABACK (1973) "Why do sickle screening in children?" Pediatrics 51 (April).

GOLDSTEIN, M. S., S. GREENWALD, T. NATHAN, F. MASSARIK, and M. M. KABACK (1977) "Health behavior and genetic screening for carriers of Tay-Sachs disease: a prospective study." Social Science and Medicine 11.

GOODMAN, R. M. (1979) Genetic Disorders Among the Jewish People. Baltimore: Johns Hopkins Press.

GREENWALD, S. et al. (1975) A Tay-Sachs Disease Prevention Program Manual. Torrance, CA: Harbor General Hospital.

Group for the Advancement of Psychiatry (1970) The Right to Abortion: A Psychiatric View. New York: Charles Scribner.

HENDIN, D. and J. MARKS (1978) The Genetic Connection. New York: William Morrow.

KABACK, M. M. [ed.] (1977a) Tay-Sachs Disease: Screening and Prevention. New York: Alan R. Liss.

--- [ed.] (1977b) Progress in Clinical and Biological Research, Volume 18. New York: Alan R. Liss.

--- and ZIEGER

KABACK, M. M., M. H. BECKER, and M. V. RUTH (1975) "Sociologic studies in human genetics I. Compliance factors in a voluntary heterozygote screening program," in Birth Defects, Washington, DC: March of Dimes National Foundation.

LEVY, H. L. (1973) "Genetic screening for inborn errors of metabolism." DHEW publication HSA 75-5708, Washington, D.C.

LITTLEFIELD, J. W. (1973) "Recent experience with prenatal genetic diagnosis," in Genetics and the Perinatal Patient. Evansville, IN: Mead Johnson.

McQUEEN, D. V. (n.d.) "Social aspects of genetic screening for Tay-Sachs disease: the pilot community screening program in Baltimore and Washington." Social Biology 22, 2.

MARGOLIN, C. R. (1978) "Attitudes toward control and elimination of genetic defects." Social Biology 25 (Spring): 33-37.

MARSTON, M. V. (1970) "Compliance with medical regimens: a review of the literature." Nursing Research 19 (July/August).

MASSARIK, F. (1980) "Socio-ideological differentiation in the U.S. Jewish population," pp. 143-162 in U. O. Schmelz et al. (eds.) Papers in Jewish Demography. Jerusalem: n.p.

--- (1974-1975) "National Jewish population study: a new U.S. estimate," pp. 296-302 in American Jewish Yearbook. New York: American Jewish Committee.

--- (1974) "Jewish identity," in CJFWF, National Jewish Population Study. New York: Author.

--- and A. CHENKIN (1973) "United States national Jewish population study," pp. 264-315 in American Jewish Yearbook. New York: American Jewish Committee.

MASSARIK, F., M. M. KABACK, S. GREENWALD, T. J. NATHAN, M. ROSENTHAL, and D. M. BASS (1977) "Community based genetic education, communication channels and knowledge of Tay-Sachs disease," pp. 353-366 in M. M. Kaback (ed.) Tay-Sachs Disease Screening and Prevention. New York: Alan R. Liss.

MILUNSKY, A. (1979) Know Your Genes. New York: Avon.

National Institutes of Health (1979) "Predictors of hereditary disease or congenital defects." Task Force Report for the Consensus Development Conference on Antenatal Diagnosis, U.S. Department of Health, Education and Welfare, Bethesda, Maryland.

OMENN, G. S. (1978) "Prenatal diagnosis of genetic disorders." Science 200 (May 26).

REAM, A. K. (1978) "Advances in medical instrumentation." Science 200 (May 26).

ROSENSTOCK, I. M. (1966) Article in Milbank Memorial Fund Quarterly 44: 94-127.

Science (1977) "Gene-splicing: Cambridge citizens okay research but want more safety." Volume 195 (January): 268-269.

SCRIVER, C. R. et al. (1978) "Genetics and medicine: an evolving relationship." Science 200 (May 26).

SEASHORE, M. R., C. AUSTEIN, and S. S. MICK (1976) "Should Tay-Sachs screening be conducted in unmarried college students?" Pediatric Research 10, 349.

SPANIER, G. B. (1976) "Use of recall data in survey research on sexual behavior." Social Biology 23 (Fall): 244-253.

SPERBER, M. A. and L. JARVIK [eds.] (1976) Psychiatry and Genetics: Psychosocial, Ethical, and Legal Considerations. New York: Basic Books.

WADE, F. C. (1975) "Potentiality in the abortion discussion." Rev. of Metaphysics 29 (December): 239-255.

AUTHOR INDEX

SUBJECT INDEX

Manageability 47, 50-53

ABOUT THE AUTHORS

FRED MASSARIK is Professor of Behavioral Science in the Graduate School of Management at University of California at Los Angeles. He is also Chairperson of the Behavioral and Organizational Science Faculty in the Graduate School of Management and is editor of the journal *Small Group Behavior.* He also serves as Director of the U.S. National Jewish Population Study. Since receiving his Ph.D. in psychology at UCLA in 1957, Massarik has authored books and articles on a wide range of social and behavioral topics including *Leadership and Organization* (with R. Tannenbaum and I. R. Weschler), *Mathematical Explorations in Behavioral Science* (with P. Ratoosh), and *The Course of Human Life* (with C. Buhler).

MICHAEL M. KABACK, M.D. is Professor in the Departments of Pediatrics and Medicine at the UCLA School of Medicine and Associate Chief of the Division of Medical Genetics, Harbor UCLA Medical Center. He has played a principal role in the development of the California Tay-Sachs Disease Prevention Program and in the development of other genetic disease programs. Kaback received his M.D. from University of Pennsylvania in 1963. He has been honored by a B'nai B'rith Service Award (Philadelphia, 1971) and was a Joseph P. Kennedy Jr. Memorial Foundation Research scholar from 1968-1972. Kaback is author of numerous publications relating to detection and prevention of genetic diseases including his edited volume, *Tay-Sachs Disease: Screening and Prevention* (1977).